Fidgeting Fat, Exploding Meat & Gobbling Whirly Birds

and other delicious science moments

KARL KRUSZELNICKI

Illustrations by Karen Young and Adam Yazxhi

JOHN WILEY & SONS, INC.

New York • Chichester • Weinheim • Brisbane • Singapore • Toronto

First published in Australia by HarperCollins Publishers.

First U.S. edition published by John Wiley & Sons, Inc.

This publication is designed to provide accurate and authoritative information in regard to the subject matter covered. It is sold with the understanding that the publisher is not engaged in rendering professional services. If professional advice or other expert assistance is required, the services of a competent professional person should be sought.

Library of Congress Cataloging-in-Publication Data:

Kruszelnicki, Karl
    Fidgeting fat, exploding meat, and gobbling whirly birds and other delicious science
  moments / Karl Kruszelnicki. — 1st U.S. ed.
      p.   cm.
    Includes bibliographical references.
    ISBN 0-471-38118-7 (pbk. : alk.)
    1. Science—Popular works.   I. Title.

Q162 .K775   2000
500—dc21   99-086955

Printed in the United States of America

10 9 8 7 6 5 4 3 2 1

THANKS

First, I thank the scientists, engineers and medical doctors who did the original research, especially the ones who had time to help me — Ed Zanotto and Yvonne Stokes ("Glass: Flowing Liquid Stronger Than Steel?"), Morse Solomon ("Explosives Tenderise Meat") and Bill Hammer ("Gobbling Whirly Birds").

Second, I thank my family (Mary, Little-But-Not-for-Much-Longer Karl, Alice and Baby-Who-Needs-No-Sleep Lola Scarlett) for reading the stories and improving them.

Third, I apologise to the staff at HarperCollins Australia for increasing their workload by being nearly three (3!!) months late in delivering the manuscript — the wonderfully patient Jane Morrow (editor), who found beautiful short cuts and neat phrases, Alison "Ali-the-Axe" Urquhart (calmest of commissioning editors), Luke Causby (book designer), Peter Guo (typesetter) and Kate Thomas (most effective publicity agent in the Entire Known Universe). Thanks also to Karen-the-Happy-Giggler Young and Adam-the-Cool Yazxhi for the great illustrations.

Fourth, I again thank Dan-the-Man-With-Street-Cred Driscoll (ABC Radio) for polishing my radio stories into shining gems, and Jason Moore, Woo Wei and Brian Fletcher for their comments and punchlines.

Fifth, I thank the webmasters of our more-than-1.5 million words homepage (www.abc.net.au/science/k2) at the Lab — Ian Allen, Alf Conlon, Deb O'Brien and Damon Shorter.

Sixth, I thank my agent — the wunnerful ex-University-Literature-Lecturer-With-a-Ph.D. Rosemary Creswell, and her colleague Greatest-Undiscovered-Singing-Talent Annette Hughes.

Finally, I thank Caroline Helldriver Pegram. Her incredible tracking and data acquisition skills, her calmness and humility, and her ability to think outside the square to juxtapose non-linear post-modernist parallel paradigms that sharpen the saw in a win-win situation make it possible for us to visit schools, write books, give talks, deliver *New Scientist* magazines to nearly 200 schools around Australia and field 20 000 inquiries each year.

And really finally, I thank all the people who found "science" mistakes in my previous books. I repeat the offer of a free autographed copy of my next book for the first detection of any "science" mistakes in this book.

KARL KRUSZELNICKI

FIDGETING FADES FAT

In another book (Pigeon Poo, the Universe & Car Paint), I touched on this topic of losing weight by fidgeting. Since then, more research has been done. I found it so fascinating that I had to write not just two paragraphs, but an entire story.

One third of the world eats too much, while two thirds of the world doesn't get enough food to eat. It's really ironic that in the wealthy parts of the world there has been a recent epidemic of obesity. But a few scientists have shown that some people can fidget their fat away.

The Balance

People have thought about body weight for a long time. Thousands of years ago, the Greek doctor Hippocrates said: "*do not allow the body to attain extreme thinness, for that too is treacherous. But bring (the body) only to a condition that will naturally continue unchanged, whatever that may be.*"

In general, there is a balance between how much food (energy) you eat and how much energy you burn up. If the balance shifts, you will get heavier or lighter.

As people age from 20 to 50 years, they tend to double the amount of fat that they carry. One reason for this could be that they may eat more and exercise less. Another reason is that your body's ability to break down and then use fat decreases as you age. So you store more fat instead of burning it up. Occasionally, hormonal disorders (hypothyroidism, or tumours of the adrenal or pituitary glands) can also cause obesity.

Fat Gives You Diseases

We now know that in some people, being too fat can cause a whole bunch of different diseases. Fatness can bring on anything from varicose veins to diabetes, osteoarthritis to endometrial cancer, infertility to gall bladder disease. It can also cause high blood pressure and heart disease.

But fatness is not always related to heart disease. For example, the Eskimos eat more fat than anybody else, yet they hardly ever suffer from heart diseases. This is because the fats they eat (fish, seal and whale blubber) are unsaturated. In general, it's the saturated fats that help cause heart diseases.

"Overweight" means that you are carrying a bit of flab, while "obese" means that you are carrying way too much fat. In Australia in 1999, about 18% of adults measured up as obese — compared to 9% back in 1989. In the United States, the situation is worse — about half the adults are overweight, and about half of those are obese.

It's easy to talk about being "overweight", but how do you actually measure it?

How to Measure Fat — BMI

The nutrition scientists have a measure called the "BMI", or Body Mass Index. This equals your weight in kilograms, divided by the square of your height in metres.

A BMI under 25 puts you in the low health risk category. A BMI between 26 and 28 gives you a moderate health risk. The health risks increase very rapidly as your BMI climbs over 30. Just for the record, obesity is defined as a BMI of over 30.

BMI Isn't Everything

But like everything related to the human body, it's not that simple.

A 70-kg (154-lb) person could be fit and lean with a lot of muscle, or they could be unfit, soft and flabby. The fit lean

BMR = BASAL METABOLIC RATE

The **BMR** (Basal Metabolic Rate) is a measure of how much energy an animal needs just to keep itself alive.

There are a few specific conditions you need to set up when you want to measure someone's **BMR**. The person has to be resting quietly, and should have slept for at least 8 hours. Their last exercise (e.g., walking) should have been at least 30 minutes earlier, and their last meal should have been at least 12 hours earlier.

Under these conditions, the **BMR** measures the minimum amount of energy needed to keep someone alive. This energy is spent moving air in and out of the lungs, blood into and out of the heart, and food along the gut. It is also spent maintaining muscle tone and electrical activity in the cells, normal background manufacture of various hormones, normal operation of various pumps in your cell membranes, and so on.

When you're resting, about 50% of your energy powers your muscles (lungs, heart, skeletal), while 20% powers your brain. Another 20% is used to power your liver, and 7% for your kidneys. The remaining 3% is used in little bits and pieces throughout your body (e.g., nerves).

person with muscles has, on average, a greater life expectancy than the fat flabby person.

In fact, a fit fat person is better off than an unfit skinny person. Chong Lee, a statistician from the University of Birmingham in Alabama, looked at 21 856 men with various weights and body shapes. According to the report in *Science*, his team found that "*unfit, lean men with BMIs of 25 or less had twice the risk of mortality from all causes than fit overweight men with BMIs of 27.8 or more*".

And of course, people with different genetic backgrounds could have different "ideal" BMIs.

Fat Belly Worse than Fat Bottom

It turns out that a fat belly is more dangerous than a fat bottom and thighs. So the pear shape is healthier than the apple shape. Carrying fat on your belly is riskier for your health, especially for heart disease and mature-onset diabetes.

There are at least two quite different mechanisms involved in the increased health risk of a fat belly.

First, it turns out that the fat cells in the belly are very good at turning fat into fatty acids, which then get turned into glucose. This flood of glucose overwhelms some cells, so they can't absorb it all. So then the blood glucose level goes up, and this can sometimes set you up for diabetes.

Second, fat cells in the belly are very sensitive to cortisol — one of the hormones involved in stress. This can cause a condition where people with lots of fat in their belly get the cortisol in their blood spiking up to very high levels during the day. Cortisol has many effects. It can promote the absorption of fat, it can make your body resistant to insulin (and again set the scene for diabetes), and it can also make your body start producing the fats that damage your heart.

Fat Costs

According to the Institute of Medicine in the United States, fat people cost the United States economy more than $US70 billion each year — both in *direct* health care costs (e.g., treating diabetes and heart disease) as well as in *indirect* costs (e.g., lost productivity). Overweight Americans spend another $US40 billion a year on trying to lose weight.

Australia has a smaller population, so the costs involved are smaller. Obesity costs the Australian health care system some $A830 million each year, while Australians spend over $A500 million a year trying to lose weight.

Early Proof That Fat Is Bad

The first real evidence that being overweight could be bad for you came from the Metropolitan Life Insurance Company in America. In 1959, they published their Risk Tables, after analysing the health of hundreds of thousands of their insurance policy holders. The Risk Tables showed that if your weight was higher than a so-called "desirable weight", you had a greater chance of dying early. This "desirable weight" was actually quite lean — 57 kg (126 lb) for a 1.63 m (5 ft 4 in.) tall woman, or 70 kg (154 lb) for a 1.78 m (5 ft 10 in.) tall man. About 80% of American men and woman are currently heavier than this "desirable weight".

ANOTHER WAY TO BURN FAT

Most of the fat in your body is stored as "white" fat. But some fat is stored as "brown" fat. Brown fat cells are quite different from white fat cells — for one thing, they have a very rich nerve supply.

There is another difference.

White fat is turned into fatty acids, which are a fuel for muscles. Muscles then shiver, and generate heat. But babies can't shiver. They have a special biochemical pathway that turns brown fat directly into heat. Hibernating animals also burn brown fat to warm themselves as they wake up. This process is called "adaptive thermogenesis".

When some animals get cold, they increase the activity in their brown fat cells. Bruce M. Spiegelman from the Dana-Farber Cancer Institute and Harvard Medical School in Boston studied mice. He and his team kept mice at refrigerator temperatures for 3 hours. They found that the mice increased their brown fat activity by 30–50 times to warm themselves. According to a report in *Science News*, he said: *"It's already been established that when animals are deficient in thermogenesis, they get fat. When people talk about having a fast or slow metabolism, this is actually what they mean."*

So this is yet another theory about why some people can eat lots but not get fat. They can store fat in brown fat cells (not the white fat cells) and just "burn" up this brown fat.

It may be possible, one day, to artificially increase this "adaptive thermogenesis" in overweight people to help reduce their weight.

In 1983, the results of the Framingham Study were published. This study looked at 5209 men and women, between 30 and 60 years of age, who lived in Framingham, Massachusetts. It analysed many aspects of their lives over a 26-year period. The Framingham Study showed that the more you swelled above your "desirable weight", so did your chances of heart disease and an early death.

In fact, the health risks increased even if you were only slightly overweight. Robert Garrison specifically analysed just the men in the Framingham Study. He found that if they were only 20% above their "desirable weight", they still had an increased early death rate from many different causes.

Many other studies backed up the finding that being fat can be bad for your health.

Recent Proof That Fat Is Bad

In 1995, the Nurses' Health Study was published. It had followed more than 115 000 female nurses for about 15 years. Its results were similar to those found by the Framingham Study and the Metropolitan Life Insurance Company. The Nurses'

Health Study found that "overweight" nurses tended to die earlier and suffer from more cardiovascular diseases.

The Nurses' Health Study found that very skinny women with a BMI of less than 19 were in the lowest risk group. Once the nurses had a BMI between 19 and 25, their health risks increased by 20%. And then the curve of BMI/risks accelerated upward — a BMI of 27–28.9 increased their risks by 60%, while a BMI of 29 or higher doubled their health risks.

But Fat Is Not Always Bad . . .

It gets more complicated. For example, the link between an "increased risk of early death" and a "high BMI" does not seem to hold for people over the age of 60 or so. We have no idea why.

To add to the confusion, some people can be "overweight" all their lives and not suffer bad effects. Once again, we don't know why.

The Take-Home Message

But in general, carrying fat is bad for your health, especially if the fat is around your belly. And the fatter you are, the greater the health risk. At the moment, the recommended maximum waist measurements are 100 cm (39.4 in.) for men, and 95 cm (37.4 in.) for women.

If you go from overweight to "normal" weight, do your health risks improve? Probably, but we haven't proved it yet. Losing weight can lower your blood pressure, and the levels of sugar and various fats in your blood. But we have not yet done the studies to tell us if losing weight lowers the rate of heart disease and early death.

Science of Fat

One thing we do know is that losing weight is hard for some of us. We also know that some lucky people seem to be able to fidget their weight away. To understand why, we need to understand fat.

So I'll chew the fat about fat.

100% DEATH RATE

It's silly to talk about an "increased death rate". We all eventually die. As the old medical joke goes, *"Life is a sexually transmitted disease, and the death rate is 100%"*. You can't get a death rate higher than that.

So why do you still get reports in radio, TV and newspapers saying that a certain activity (such as eating fatty foods, being a pedestrian, skydiving, etc.) "increases your death rate"? In many cases, what they meant to talk about was an "increased early death rate" — dying before the average life span for your social group. These life spans vary enormously from country to country. On average, Japanese women living in Japan live for about 82 years, while men in Ethiopia live for just 39 years.

THE FAT INDUSTRY

Over 90% of all the fat used by humans comes from just 20 species of plants and animals. Most of this fat is eaten, but some is used in industry.

The ancient Egyptians used olive oil to lubricate building materials like heavy rocks, and 3200 years ago they used fat to make grease for their axles. Homer, the ancient Greek poet, mentioned oils being used to make weaving easier. Fats were also used to make soap, candles and cosmetics.

More recently, Napoleon needed a substitute for butter, so he offered a prize for a new invention. In 1869, a French chemist, Hippolyte Mège-Mouriès, won this prize when he created margarine.

Today, most edible fats are classified as liquid oils (coconut, olive, peanut, corn, soybean, etc.) and plastic fats (butter, margarine, lard, etc.).

The average amount of fat you eat each year depends on your culture. It varies from 25 kg (55 lb) in the wealthy countries, to a worldwide average of 10 kg (22 lb), and down to 5 kg (11 lb) or less in Southeast Asia, Africa and South America.

Fat Stores Energy

Fat is more than just a great insulator against the cold. It is a great way to store energy and is often called "Nature's Storehouse of Energy". Fat repels water — so you can store energy without having to store water. Weight for weight, fat stores about twice as much energy as carbohydrates and proteins. Fat is broken down into fatty acids, which can then be used directly by the muscles as a fuel.

In vertebrates (animals with spines), about half the energy used by organs such as the heart, kidneys and liver comes from burning fat. Fat supplies almost all the energy in migrating birds and hibernating animals. Flying insects also get most of their energy from fat.

Fat is stored in special cells called "apidocytes". In humans, fat is stored mostly in the adipose tissue just under the skin, but there are also various deposits of fat scattered around the body. It is often found wrapped around the heart, in between the muscles and near the small and large intestine. This fat is continually being broken down and then restocked, even though your total amount of fat may stay the same.

Fat Can Keep You Alive for Months

The main advantage of carrying a bit of fat is that it will see you through lean times. The average American male carries enough fat (15 kg or 33 lb) to meet his energy needs for two months. An obese person could carry 110 kg (220 lb) of fat, which is enough to supply their energy for a whole year!

You have to have a certain minimum amount of fat. When a man is as lean as he can possibly be, he is still carrying about 3 kg (6.6 lb) of fat.

Men tend to carry their fat around the belly. This fat can easily be turned into fuel for a quick energy hit. But women tend to carry their fat around the thighs and buttocks. This fat tends to be the energy source to grow the baby in pregnancy, and to make breast milk once the baby is born.

People Handle Fat Differently

Why does eating the same amount of food make some people fat and leave other people skinny?

Back in 1902, a scientist called Neumann realised, after he had experimented on himself, that this was a real phenomenon, so he wrote it up. By 1972, another scientist called Sims overfed some volunteers and realised that some people were "easy gainers" (put on weight easily), while others were "hard gainers" (found it hard to put on weight).

Early Fidgeting Research

Professor Leonard Storlein, from the Department of Biomedical Science at the University of Wollongong, accidentally discovered the weight loss benefits of fidgeting. He was measuring the Basal Metabolic Rates (BMR) of various humans. The BMR is a measure of how much energy you need just to keep your body running, while you're hanging around, doing nothing.

He used a device called a Whole-Room Calorimeter. A whole-room calorimeter is just a small, sealed room. Because the room was sealed, he could measure the total amount of oxygen his volunteers used, and how much carbon dioxide they produced. From this, he could work out how much energy his volunteers were burning up.

Depending on your size, your BMR is between 1500–2000 Cal (6280–8368 kJ) per day. To his surprise, he found that some people were burning up extra energy above the expected BMR, depending on how much they fidgeted. The extra energy varied from 200 Cal (837 kJ) per day for somebody who just lounged around to 1200 Cal (5021 kJ) per day for a dedicated fidgeter. The difference of 1000 Cal (4184 kJ) per day is an amazingly large amount.

Fidgeting Keeps You Skinny

According to Professor Storlein, "*A person would normally run 10 km just to get rid of 1250 kJ*". So 1000 Cal (4184 kJ) is roughly

FAT PLANTS

Most of the fats in plants and nuts are the "good" unsaturated fats.

Most plants are low in fat (0.1–2%). But a few plants, such as olives and avocados, have lots of it (70%).

In general, the amount of fat increases as the plant gets closer to maturing the seed. The fat is there to nourish the seed as it grows in its new home.

Cereals contain between 1–7% fat, while nuts can have up to 70%.

HOW LONG CAN YOU FAST?

According to Western medicine, on average, you can survive two months without food, or 72 hours without liquid. In 1981, Bobby Sands, the IRA (Irish Republican Army) hunger striker, finally died after 66 days without food.

But according to the Breatharians, a 5000-strong group, you don't really need to eat at all. The *Fortean Times* reported that all Breatharians need is fresh air and sunshine.

However, back in 1983, their then-leader, 47-year-old Wiley Brooks, sneaked into a hotel to order a chicken pie — and was caught in the act. He, and most of his followers at the Breatharian Institute in California, had to resign.

In Brisbane, on the 2 July 1998, Lani Morris died after attempting the Breatharian initiation — seven days without liquid, 21 days without food. Her Breatharian mentors were charged with manslaughter. One of them, Jim Vadim Pesnak, claimed not to have eaten for three years.

Another Brisbane Breatharian is almost unknown in Australia, but is "famous" in Europe and the United States for her eight best-selling books. Her latest book is *Living on Light*. Ellen Greve, aged 41, an ex-bank clerk with Norwegian parents, now calls herself Jasmuheen. She claims that she hasn't eaten for five years. Instead, she has trained her body to live on '*prana*', or "liquid light". It seems that her understanding is different from the traditional yoga interpretation of *prana* as "breath, life, respiration, vitality, energy, wind or strength".

Thanks to her claimed high level of "spiritual enlightenment", not only can she talk telepathically with the "Ascended Masters", but angels will also organise good parking spots for her car.

"Gosh I'm full"

THE FATTEST BIRD

When birds migrate long distances, they get the energy to fly from burning up fat. Your average non-migrating bird carries only 3–5% of its body weight as fat. Birds that migrate short distances carry 20–30% fat, while some long-haul birds can reach 40–50% fat.

But the Bar-Tailed Godwit (a shorebird) holds the record at 55% fat! Because they are so fat, they can store enough energy to make them the long-distance marathon champions of flight — 11 000 km (6800 miles) non-stop across the Pacific.

To prepare for the flight, they do more than just pig out and lay down fat. They also shrink a few organs that are not needed for the migration, such as their liver and stomach. According to Theunis Piersma from the University of Gröningen and the Netherlands Institute for Sea Research, *"Everything you don't really need for the flight is thrown overboard"*. The bar-tailed godwits probably undergo other profound changes in physiology that we haven't discovered yet.

Providing they can lumber down the runway and get their obese bodies airborne, not only do they get across the Pacific in one hop, they even arrive for their holiday slimmed down to half their take-off weight.

equivalent to a 33-km run. In other words, your dedicated fidgeter burns up enough energy to run 33 km (20 miles) each day!

My brother-in-law James is as skinny as a rake and eats like a horse. He is a serious fidgeter. He twiddles his thumbs, bobs up and down, and continually crosses and uncrosses his legs. He also continually taps his feet. I guess that moving the long muscles of the body (e.g., leg muscles) burns up more fat than moving the short muscles (e.g., fingers). And of course, most fidgeters do it all day long.

Fidgeting has another benefit. It tones up your muscles.

However, we have not done a lot of research on the Finer Points of Fidgeting. For example, we don't know what type of fidgeting burns the most fat.

PENGUINS LIVE OFF FAT

Down in the Antarctic, it's the male Emperor Penguin who hatches the single egg. He sits on the egg during the dark polar winter for 105–115 days. He can't leave the egg or else it will freeze to death, so he has to live off his fat stores.

Thousands of the male Emperor penguins all huddle together with their backs to the wind. Yvon Le Maho and her team from the Centre National de la Recherche Scientifique in Strasbourg, France, studied the efficiency of the huddle. They compared eight penguins that were in a huddle of 3000 other penguins with 10 solitary penguins. The huddling penguins burned up their energy 17% more slowly than the solitary penguins and by the end of winter weighed 7% more.

Life is finely balanced in the Antarctic. While the huddling penguins will survive with their eggs, solitary penguins will run out of energy three weeks before the eggs hatch and will have to abandon their eggs to get food.

SKINNY SINGAPORE SOLDIERS

In Singapore, 18-year-olds have to serve two years of National Service. But the two-year clock doesn't start ticking until the soldier is down to the Army's maximum allowable weight.

So one quarter of all the incoming soldiers have to spend an extra four months in the Army, to get their weight down to the Army's maximum weight.

More Fidgeting Research

Just recently, Doctors James Levine, Michael Jensen and Norman Eberhardt from the Mayo Clinic in Rochester, Minnesota, did some more "fidgeting" research .

They began with 16 volunteers (12 men and four women) who were all healthy, not obese, and aged between 25 and 36. The volunteers weighed 53–92 kg (117–203 lb), and had BMRs between 1500–2000 Cal (6280–8368 kJ) per day. On average, they needed another 1100 Cal (4600 kJ) to do normal everyday activities, such as walking around, scratching yourself, brushing your teeth and walking over to open the refrigerator door.

The experiment was to overfeed the volunteers by giving them an extra 1000 Cal (4184 kJ) per day on top of what they usually ate.

No Cheating Allowed

The doctors supervised the volunteers quite closely for eight weeks. As the volunteers got fatter, the doctors were very thorough in measuring the increasing fat content of their bodies by using a process called "Dual Energy X-Ray Absorptiometry". They also made sure that the volunteers didn't sneak in any extra exercise, by getting them to wear pedometers (a pedometer is roughly the size of a watch, hangs from your belt, and clicks forward one notch every time you take a step).

They checked with the friends and family of the volunteers to see if the volunteers were not eating their meals, or sneaking off to the gym. The scientists also measured the amount of energy the volunteers used with "Doubly Labelled Water". They looked in their garbage bins to make sure they weren't throwing away the extra food. The doctors even sometimes analysed the volunteers' faeces, just to make sure that they were following the "No Exercise, Eat More" plan.

Fidgeting Results

The doctors found a surprising variation in weight gain, from 1.4–7.2 kg (3.1–15.9 lb).

Why did some volunteers put on five times more weight than others?

The answer lies in what happened to the extra food — the extra 1000 Cal (4184 kJ) per day that they ate.

Where the Energy in the Food Went

About 8% of the energy in the food was spent increasing their BMR. Another 14% of the energy in the extra food was used up absorbing, digesting and storing that extra food.

The big difference between volunteers appeared in what the doctors called "NEAT" (NonExercise Activity Thermogenesis). In plain English, NEAT is just random activity which is not specifically exercise — what you and I would call fidgeting. The amount of energy spent in these non-exercise activities varied between 10% (easy gainers of weight) and 50% (hard gainers of weight).

A hummingbird has an incredibly high metabolic rate. Its heart can beat 1200 times each minute, so it has to eat huge amounts of food.

On a very good day, a hummingbird can eat three times its own body weight. About one third of their diet is insects, while the rest is "sugar water", that is, nectar from flowers.

The remaining 25 to 65% of the energy in the food got turned into fat.

Problems with This Study

Now there were a few problems with this study — but of course, this was just an early study. First, even though the volunteers all had to eat an extra 1000 Cal (4184 kJ) per day, their starting weights varied enormously. So the extra 1000 Cal (4184 kJ) per day wasn't graded for their starting weight (53–92 kg or 117–203 lb). Second, the sample size (12 men and four women) was very small.

Even so, it *is* an interesting result, because it gives us an idea that we could fidget our way to slimness. Perhaps we could even introduce Fidgerobics as a really NEAT sport into the Olympic Games.

REFERENCES

"Fat-Fighting Fidgeting", *Scientific American*, March 1999, p 16.

Fortean Times, March 1999, p 14.

James A. Levine, Norman L. Eberhardt, Michael D. Jensen, "Role of Nonexercise Activity Thermogenesis in Resistance to Fat Gain in Humans", *Science*, Vol. 283, 8 January 1999, pp 212–214.

"Obese Birds Make Good Athletes", *Science News*, Vol. 153, No. 2, 10 January 1998, p 28.

Eric Ravussin and Elliot Danforth Jr., "Beyond Sloth — Physical Activity and Weight Gain", *Science*, Vol. 283, 8 January 1999, pp 1, 184–185.

Professor Ben Selinger, *Chemistry in the Market Place*, Harcourt Brace & Company, 1998, pp 90–93, 107–108.

Ingrid Wickelgren, "Obesity: How Big a Problem?", *Science*, Vol. 280, 29 May 1998, pp 1364–1367.

MATHEMATIC ADDICT

Paul Erdös (pronounced "AIR-dish") was one of the greatest, most prolific and most original mathematicians of all time. He slept for only three hours a night, and did mathematics seven days a week, 19 hours a day until he died at the age of 83. He believed that a mathematician was a device for turning coffee into mathematical theorems.

Physical Is Nothing ...

Erdös loved only mathematics. He once said, *"I cannot stand sexual pleasure. It's peculiar."* He didn't care about property, food, clothes or paying taxes. He never learnt how to prepare or cook food, although he could add milk to breakfast cereal. His only possessions were some old clothes and a couple of battered suitcases.

Erdös followed the old Greek saying: *"The wise man has nothing he cannot carry in his hands".*

He crisscrossed the world as an itinerant gipsy mathematician. He never stayed anywhere for more than a month. Erdös was like a bee flitting from flower to flower. But while a bee carries out sexual cross-pollination for flowers, Erdös would pick

up new problems and drop off results from other mathematicians — a kind of mathematical cross-pollination.

The deal was simple. His fellow mathematicians would take care of his physical needs, and he would bless them with the fruits of his brain.

Taking Care of Paul

Part of the price of having Paul Erdös and his brain visit you was having to do an "Uncle Paul Sitting". This meant taking care of Paul's current physical needs: getting him some money; organising transport details to his next port of call; or driving him to the doctor to get more amphetamine tablets.

Erdös' social skills were virtually non-existent. It seems he couldn't even shut doors or windows. On one occasion, his hosts had left a window open for him so that he could climb into the house when he arrived a little after midnight. Later that morning, before the sun had risen, a very heavy rain began flooding the ground floor of the house. Erdös didn't shut the window. Instead, he woke his friends and said: "*It's raining in the window. You'd better do something.*"

Paul was always supremely confident that everything would work out fine. For example, in 1984 he left America to go to Japan with only $50 in cash — no cheque book and no credit card. But, as always, he believed that the combined forces of mathematicians in 25 countries on four continents would take care of him. "*There was no reason to worry. I had friends everywhere along the way.*" He was correct. For mathematicians, he was a National Living Treasure.

Erdös' Gift to Mathematics

It was a great honour for a mathematician to be invaded by Erdös.

He would suddenly turn up at the front door of one his fellow mathematicians in any of 25 countries, and announce that he was ready to do maths by saying, "*My brain is open*". He would stay for up to a month, until he had exhausted his host. Erdös would fuel his brain with coffee, caffeine tablets and Benzedrine. They would work until the early hours of the morning, when the exhausted host would eventually stumble off to bed.

A few hours later, the host would be woken by the inquiry "*Do you exist?*" which was Erdös' way of asking if he or she was awake. Then, without waiting for an answer, Erdös would immediately launch into "*Let n be an integer and k be a set such that . . .*" On another occasion, he burst into a Christmas party and button-holed his host with "*Merry Christmas. Let f of n be the following function . . .*"

Even so, the mathematicians loved to be invaded, because of the exhilarating time they had with Erdös while they danced together through the Land of Mathematics. While sailors might have the motto "*A wife in every port*", Erdös had the motto "*Another roof, another proof*".

Erdös refused to slow down, saying "*There'll be plenty of time to rest in the grave*".

Paul Was Wide and Deep

Paul Erdös was very different from other mathematicians.

Practically all mathematicians spend most of their life in one field, often working

My brain is open

SPEED MAKES YOU STUPID

Stimulant drugs such as cocaine and amphetamine will increase your work output. But they also destroy your ability to tell the difference between what is good and what is rubbish. You might produce more, but you usually produce rubbish.

The history of the modern music industry is littered with the bones of hundreds of bands that have taken stimulant drugs in the course of writing their music. While they were under the influence of the stimulants, the music they wrote and performed sounded fantastic. But to everybody else, and to themselves when they stopped taking the stimulants, it sounded like complete garbage. Several famous bands have written albums while high on stimulants, and then have had to dry out from the drugs, and start again from the beginning.

Erdös was one of the incredibly rare people upon whom stimulants actually had a positive effect. He specifically warned other mathematicians against taking them, because he admitted that his was a special case.

RECENT MATHEMATICS

Most people would be ashamed to admit that they cannot read or write. But those same people will be proud of their inability to do mathematics and say openly *"Oh, that's too hard for me, I can't do maths"*.

In literature, the average well-educated person would certainly have read a book written in the last few years. In science, the average well-educated person would have heard of the Big Bang theory in astronomy, or of DNA in genetics, or the relationship between CFCs and the hole in the ozone layer.

But in mathematics, very few people would know anything more recent than what was discovered since the Middle Ages. They know a little bit of geometry (from the ancient Greeks), the zero (from the early Muslim mathematicians) and a little algebra (from the European Renaissance). Hardly any "well-educated" or "well-rounded" people would know calculus, which was invented in the 1600s.

Most people would not know any mathematics produced more recently than 500 years ago. Oddly enough, this doesn't seem to bother them.

on a single problem. For example, Andrew Wiles spent the best part of a decade solving Fermat's Last Theorem. He consulted with very few fellow mathematicians. But Erdös delighted in working with other mathematicians, and worked in many different fields at the same time.

He was one of the most prolific mathematicians in history. A typical first-class mathematician might publish up to 100 papers in an entire lifetime. Each paper would have new and original concepts that have previously never been published. But Erdös wrote an amazing 1500 papers — roughly one paper every 15 days.

Erdös was skilled in many areas of mathematics. He pioneered the fields of Number Theory and Combinatorics. Combinatorics is the very foundation of computer science. He founded the field of Probabilistic Number Theory with Mark Kac and Aurel Wintner. He worked with Paul Turan doing major work in Approximation Theory.

In 1949, Erdös astonished fellow mathematicians by working with Atle Selberg to come up with an elementary proof of the Prime Number Theorem. The previous proof worked, but it was *"rather messy"* because it used the square root of negative numbers. Their proof was much more beautiful.

Erdös and Selberg had agreed to share the credit by publishing in the same journal at the same time. But Selberg broke the agreement by rushing into print first, so he got most of the credit. Most people would be furious about this. But Erdös didn't care and continued with his work.

There's an old saying in science: *"It's not the answer that gets you the Nobel Prize, it's the question"*. One of Erdös' greatest qualities, as far as other mathematicians were concerned, was the ability to ask the question that would have a very interesting answer. He also knew which mathematician, or group of mathematicians, to ask.

Special Language

Erdös had his own brand of English. He would call women *"bosses"*, and men *"slaves"*. A child would be an *"epsilon"* (because in maths, epsilon stands for a very small quantity). *"Sam"* was the United States, and *"Joe"* (as in Josef Stalin) was the Soviet Union. So the *"Sam-and-Joe Show"* was the International News. Alcohol would be *"poison"*, while anything that wasn't speech was called *"noise"*. So *"poison, bosses and noise"* meant "wine, women and song".

A *"fascist"* was anybody who was annoying, or who put an obstacle in his path. Erdös believed that God, whom he called *"S.F."* or the *"Supreme Fascist"*, had an infinitely large book that contained all of the perfect and most elegant mathematical proofs that could possibly exist. S.F. was very mean in keeping this book secret from humans. The job of mathematicians,

ENGLISH OF MATHS

Mathematicians have a strange language of their own. A theorem might be "deep", an argument against that theorem might be "striking", while a proof might be "beautiful", "elegant", "pure" or "exceptionally beautiful".

thought Erdös, was to get these proofs "*straight from the Book*".

If mathematicians stopped doing maths Erdös would say they had "*died*". But for "die" he used the phrase "*to leave*".

Erdös and Stimulants

Erdös first used amphetamines after his mother died in 1971. He started taking them as antidepressants on a doctor's advice. He then continued taking his small daily doses. A fellow mathematician, Ronald Graham, was worried that Erdös was addicted. (Graham was once mentioned in the *Guinness Book of Records* for having used the largest number ever used in a mathematical proof. In 1996, he was the Director of the Information Sciences Research Center at the AT&T Laboratories.)

So Graham offered Erdös $500 if he could give up the "speed" for a month. Erdös thought that the small doses of amphetamines were just the same as having lots of strong coffee. He accepted the challenge, but the price was that he lost his creativity. He said: "*Before, when I looked at a piece of blank paper, my mind was filled with ideas. Now all I see is a blank piece of paper.*" He even blamed Graham for "*setting mathematics back a month*". Erdös went without stimulants for a month and collected the $500, and then went back to taking his amphetamines in small daily doses.

The Beginnings

Paul Erdös was born on 26 March 1913, in Budapest, Hungary. Within a few days of his birth, his two sisters (aged three and five) died from scarlet fever. (His parents, both teachers of mathematics, always thought that his sisters were smarter than he was.) When he was one-and-a-half, his father was captured by the Russians in an attack on the Austro-Hungarian Empire and sent to Siberia for six years.

By the time little Paul was four years old, in 1917, he was already multiplying four-digit numbers together. He had to do these calculations in his head, because he didn't know how to write down numbers. But his major achievement of the year 1917 was to invent for himself negative numbers. He said to his mother, "*If you subtract 250 from 100, you get 150 below zero*".

USELESS MATHEMATICS

Mathematics has always been the "Queen of the Sciences". It is the most magnificent, highest and purest thing ever created by the human mind.

Much of it is also supposedly useless. But given enough time, most maths will find a use.

The English mathematician G.H. Hardy wrote in his 1940 book *A Mathematician's Apology*: *"I have never done anything 'useful'. No discovery of mine has made, or is likely to make, directly or indirectly, for good or ill, the least difference to the amenity of the world."*

He was wrong. Just because his special field of Number Theory had been totally useless for the previous 2000 years didn't mean that it would always stay that way. His work in Number Theory has become very important for sending your credit card number from one computer to the next. Number Theory is used for encoding and decoding secure information.

The Greeks spent a lot of time studying the strange (and "useless") curve called the ellipse. Kepler discovered 2000 years later that the curves that planets follow around the Sun are ellipses.

Bernhard Riemann, a German mathematician, decided in 1854 to build a completely artificial geometry. Under our standard geometry, much of which was given to us by the Greek mathematician Euclid, it is definitely possible to have two lines that are parallel to each other. Riemann developed a geometry where it was impossible to have two parallel lines.

This was regarded as totally useless, until Albert Einstein declared that Riemann's strange curved space is the true shape of the universe. The only reason that we think that we can have lines that are parallel is that we see such a tiny part of the universe.

His mother kept him out of school until he was a teenager, and taught him at home. She was terrified that he would pick up an infection at school and die as his sisters had.

Erdös entered university at 17. When he was around 19, he wrote his first paper and proved his first theorem. This theorem was that there will always be at least one prime number (a number divisible only by itself and 1, such as 3 or 17) between N and 2N. In plain English, there will be at least one prime number between 3 and 6, or between 400 and 800, and so on. Chebyshev was the first person to prove this theorem, but Erdös' proof was *"far more striking, and neater"*. Erdös was suddenly famous among European mathematicians. A small poem was invented to celebrate his success. It ran:

Chebyshev said it, and I say it again,
There is always a prime between N and 2N.

He was awarded his Ph.D. at 21. He went to the University of Manchester, England, in 1934 for his post-doctoral fellowship.

By the late 1930s, the prejudice against Jews was so great that he could not return to Hungary. So in 1938 he immigrated to the United States. During World War II, most of his relatives in Hungary were killed.

In the immediate Cold War period after World War II, he attracted suspicion in the United States. Erdös dressed in a strange way, and he had a thick East European accent. The United States was paranoid about Communism. On one occasion when he tried to re-enter the United States, the customs official asked him about Karl Marx (one of the founders of Communism). Erdös didn't really know much about Marx, so he replied honestly, *"I'm not competent to judge, but no doubt he was a great man"*.

As a result of this one sentence, he was not allowed into the country.

Erdös had a lot of trouble with authorities. The Soviets didn't like him, because he spent so much time in the United States. The British didn't like him, because he corresponded with a fellow mathematician in Communist China, and used many mathematical symbols in his letters. They thought that the mathematical symbols were some kind of a secret code.

He spent much of the 1950s in Israel, but was eventually allowed to re-enter the United States in the early 1960s. He wandered the world looking for mathematical problems to solve.

Not the Solitary Worker

Erdös collaborated with more mathematicians than any other mathematician in history.

For example, he collaborated with Canfield and Pomerance to create the Canfield, Erdös and Pomerance Approximation. This Approximation can be used to decode encrypted numbers, such as enormous financial transactions between banks, or just your credit card number when you buy a book over the Net.

Kevin Bacon Number or Erdös Number

In the Kevin Bacon game, you try to link the actor Kevin Bacon to any other actor in a few steps.

Mathematicians have their own "Kevin Bacon" game. They like to show off by showing how close they were to writing a paper with Erdös — they quote their Erdös Number.

For example, in the email newsletter *RISKS 20:24* the mathematician Martin Ward wrote an article called "Regular Break-ins at the Pentagon?". He wrote: *"So there are hundreds of attempts per week and*

99.95% of them fail. Let's assume about 200 attempts per week. That means, by my calculations, there are about five *successful* attempts to break into Pentagon systems every year. Sometimes, 99.95% success just isn't good enough." He signs his article with "*Martin Ward, Erdös Number: 4*"

Low Erdös Number = High Status

Erdös himself had Erdös Number Zero. There are some 500 mathematicians who have Erdös Number One — i.e., they directly wrote a paper with Erdös. There are about 5600 mathematicians with Erdös Number Two. They have written a paper with somebody who had co-written a paper with Erdös.

Some 63 Nobel Prize winners have Erdös Numbers less than Nine. All of the winners of the prestigious Fields Mathematics Medal have Erdös Numbers less than Six. Ronald Graham has an Erdös Number of One, Albert Einstein had an Erdös number of Two, while Andrew Wiles has an Erdös Number of Four.

The famous baseball player Henry L. "Hank" Aaron has an Erdös Number of Two. He and Carl Pomerance co-autographed the same baseball, when they were both getting honorary degrees from Emory University in 1995. Pomerance had previously co-authored many papers with Erdös, so giving Hank Aaron the baseball player his Erdös Number of Two.

In fact, Erdös himself made jokes about "Fractional Erdös Numbers". If a mathematician had co-authored N papers with Erdös, then they had an Erdös Number of $1/N$. In Plain English, if you had co-authored 12 papers with Erdös, your Erdös Number would be $1/12$.

Two mathematicians have Erdös Numbers less than $1/50$, and about 30 have an Erdös Number less than $1/10$.

Erdös the Generous

Erdös was an incredibly generous man. When he won the $50 000 Wolf Mathematics Prize in 1984, he kept only $750 for himself. He gave the rest of it away, mostly to encourage young people in mathematics.

Erdös would spark up interest among the students with the magic words "*I would like to mention one of my favourite problems*". This meant that he had a problem that he personally hadn't solved, but that he thought might be solvable by somebody in the audience — for money. He had begun this habit in 1954. The prize money ranged between $10 (for a fairly easy, but still tricky, problem) up to $10 000 (for what

ERDÖS-BACON NUMBER?

Actors are linked to Bacon, while mathematicians are linked to Erdös.
Very few people are linked to both Erdös and Bacon.
Dan Kleitman is one. He has a combined Erdös–Bacon Number of Three!
He has written a paper with Erdös, and appeared in *Good Will Hunting* with Minnie Driver, who acted with Bacon in *Sleepers*.

HYPATIA

Hypatia (370–415 AD) was an early Egyptian philosopher, who was also the first famous female mathematician and astronomer. She was famous for her beauty, modesty, eloquence and massive intellect, and had many students.

Unfortunately, Alexandria was at that time in a state of great tension between the Christians and the non-Christians. Hypatia was a very visible symbol of science and knowledge, which the Christians thought of as a pagan, non-Christian activity. Cyril became the Patriarch of Alexandria. His Christian followers attacked Hypatia, scraped off her skin with oyster shells and killed her. After that incident, Alexandria went into a major intellectual decline and was no longer a centre of learning.

he considered a *"hopeless problem"* that might take a few decades, or centuries, to solve). By 1984, he had offered $30 000 in prize money, and had paid out $5000.

But besides being generous with money, he was generous with his brain to several generations of young mathematicians. He always searched for, and helped, mathematically gifted young people. He would give them problems that they could solve, and would keep a close eye on them while treating them as equals.

In 1984, Willy Moser, a mathematics professor at McGill University in Montreal, said, *"There must be 100 mathematicians today who got tenure because of problems that Erdös suggested and helped them solve"*.

Erdös' End

His mother used to say, *"Paul, even someone as busy as you can never be in two places at the same time"*. Paul Erdös died in Warsaw, Poland, on 26 September 1996, after having two heart attacks. Or in his own special language, he *"left"* on the infinitely long one-way journey for which you do not need a passport.

Possibly, he is now hanging out with all the great mathematicians of the past and the future, and he can be in many places at the same time, crunching big numbers and solving some really cosmic equations.

REFERENCES

Martha Claudio, "Close Calls", *Science News*, Vol. 154, October 17 1998, p 243.

The Erdös Number Project Website — http://www.acs.oakland.edu/~grossman/erdoshp.html

Paul Hoffman, "Man of Numbers", *Discover*, July 1998, pp 118–123.

Paul Hoffman, *The Man Who Loved Only Numbers: The Story of Paul Erdös and the Search for Mathematical Truth*, Fourth Estate, 1999.

RISKS-LIST: Risks-Forum Digest, 11 March 1999, Vol. 20: Issue 24.

John Tierney, "Paul Erdös Is in Town. His Brain Is Open", *Science 84*, October 1984, pp 40–47.

FURTHER FALLING CATS

I first looked at the topic of Falling Cats back in 1989. Since then, lots of research has been done, so I have had to rewrite the story . . .

Cats are cute and cuddly — and have a very efficient self-righting mechanism. But always landing feet first doesn't explain why it's safer for a cat to fall from a 32-storey building than from an eight-storey building!

In New York, most of the people live in high-rise apartments. The combination of humans, cats, high-rise buildings, concrete pavements (and maybe birds flying past open windows) gives a ready-made environment to test the aerodynamics and impact resistance of falling cats.

High-Rise Trauma Syndrome

The High-Rise Syndrome (HRS) in falling cats was first defined by Dr Gordon W. Robinson, of the Henry Bergh Memorial Hospital in New York, back in the September 1978 issue of *Feline Practice*. At that time, his hospital was treating some 150 cases of HRS per year.

Robinson defined the syndrome as having three features. They were a bleeding nose, a split hard palate (in the roof of the mouth), and a pneumothorax (air in the "pleural cavity", the space between the lungs and the ribs).

He discussed the treatment of HRS. The cat's bleeding nose usually healed by itself. The split in the hard palate was usually treated conservatively without surgery — just soft food and antibiotics. The treatment of the pneumothorax depended on its severity. The air in the pleural cavity was removed if the cat had difficulty in breathing, otherwise no treatment was prescribed apart from "*cage rest*" in a "*quiet ward with no dogs to cause excitement*".

The greatest heights that he recorded cats surviving from were "*18 storeys onto a hard surface, 20 storeys onto shrubbery and 28 storeys onto a canopy or awning*".

Falling Cats Survive Big Falls Better

This claim comes out of a paper in the *Journal of the American Veterinary Medical Association*, written in 1987.

The cats were not deliberately thrown out of high-rise buildings by sadistic owners; they jumped out by themselves. At least that's what the owners of the cats said when they took the cats to the Department of Surgery at the Animal Medical Center at 510 E 62nd St, New York City.

Over a five-month period (4 June to 4 November 1984), 132 cats fell from a height of at least two storeys, and ended up at this hospital. It was raining cats at about one per day. Two vets at the hospital (Doctors Wayne O. Whitney and Cheryl J. Mehlhaff) examined and repaired the plunging pussies, and did some statistics.

Stats on Cats

They found that the average fall was five-and-a-half storeys (a storey being about 3.7 metres or 12 ft). The cats fell distances ranging from two to 32 storeys.

Only three of the 132 cats were witnessed in the act of falling — two when turning around on a ledge, while the third was seen jumping for an insect.

Twenty of the 132 cats were dead within 24 hours of the fall. Three were Dead On Arrival at the hospital, while another 17 were euthanised, usually because the owners could not afford the medical costs.

The statistics on the remaining 112 cats were encouraging. They all survived with mostly bone and lung injuries.

On average, only every fourth cat that fell from a two-storey building suffered a broken bone. But every cat that fell eight storeys suffered a broken bone. So for four times the height (eight storeys versus two storeys), there were roughly four times as many broken bones. (After all, four times the height means four times the gravitational energy that the falling cat has to get rid of.)

8 Storeys Very Dangerous

But as the cats fell from heights greater than eight storeys, they broke fewer bones. There was only one broken bone among all the 13 cats that fell more than nine storeys.

As far as broken bones were concerned, seven or eight storeys was the most dangerous height to fall from. The same trend followed for chest injuries.

The bottom line? It's safer for a cat to fall from 32 storeys than from eight storeys. In fact, eight storeys is the most dangerous height for a cat to fall.

Why?

Well, first of all, this has nothing to do with their self-righting ability, which gets all their four feet pointing downwards. This happens in a metre or so. It certainly doesn't explain why they can survive a 32-storey fall.

Why Cats Survive

There are three reasons why cats survive a fall: Terminal Velocity, the Strength/Size Relationship of Bones, and Body Alignment.

Survival 1: Terminal Velocity

First: Terminal Velocity, or top speed. Cats accelerate only for the first five or so storeys. After five storeys, they have reached their top speed of 100 kph (62 mph), and do not keep on accelerating. Even if they fall off Mt Everest, they won't hit the ground at faster than 100 kph (62 mph).

Why do you reach a top speed rather than continue to accelerate?

The reason you fall when you jump out of a window is because gravity sucks. If there was no atmosphere, you would just keep on accelerating, falling faster and faster. But we do have an atmosphere. Wind resistance slows you down. When the "suck of gravity" is balanced by the "resistance of the wind", you are now travelling at what is chillingly called the terminal velocity — your top speed.

This balance depends on the size and weight of the falling object. Ants fall very slowly, while elephants probably go

supersonic if given enough height. (Just kidding, they wouldn't really, but they would probably reach 500 kph or 300 mph. I await the computer simulation.) For a human being, the terminal velocity is about 200 kph (120 mph), and very few of us survive a landing at that velocity. But a cat is lighter and fluffier, and it has a terminal velocity of roughly 100 kph (62 mph).

So cats' slower terminal velocity means they have less gravitational energy to get rid of than we humans. The cats are off to a good start.

Survival 2: Strength/Size of Bones

The second reason why cats survive a fall better than we do is that, weight for weight, smaller bones are stronger. In technical terms, we are talking about the Strength/Size relationship of long bones, such as leg bones or ribs.

To make it easier to understand, let's pretend that we have a spherical animal

MUMMIFIED CATS

A few decades ago, there was a popular book called *101 Uses for a Dead Cat*. The ancient Egyptians had an extra use. In ancient Egypt the cat had great religious significance. The Egyptians would mummify dead cats, and honour them by putting them in tombs.

In the 19th century, huge numbers of amateur archaeologists were running wild across Egypt. They found so many dead cats in the tombs that they were just thrown away. Mummified cats were used for fertiliser, or even as ballast on ships.

When scientists started looking closely at X-rays of these cats, they found that the cats did not die of old age. In fact, these mummified cats were only a few months old, and were all in good health until they were strangled. The theory is that the cats were bred by priests in the temples, and then killed and mummified to be sold to temple visitors as votive offerings.

MORE STATS ON 132 FALLING CATS

Most of the 132 cats studied had some kind of injury. Emergency treatment was needed for 37% of the cats, another 30% needed non-emergency treatment, while the rest needed no intervention.

Ninety-one cats were X-rayed, and 82 were seen to have some kind of chest injury. Eighty-three of the cats were diagnosed to have a pneumothorax (air between the ribs and the lungs). Seventy-three cats suffered abnormal breathing. Fifty-three of the cats were judged to have a pneumothorax serious enough to require thoracocentesis. (The other 20 cats were probably judged to have only a "minor" pneumothorax that would resolve spontaneously.)

Only 28% of the cats could walk normally, while 43% were lame, 27% could not walk, and 2% were "paralysed". Fifty-two of the 132 cats (39%) had a total of 81 fractures in their legs. There were roughly the same number of fractures in front and back legs.

Fifty-seven per cent had some kind of injury to the face (including a bleeding nose). Lower down the list, at 17%, were cats with fractures to the dental area, or the hard palate, or to the lower jaw. The cats with fractured hard palates were treated conservatively with a moist diet and antibiotics, and all healed within a month.

(physicists like to do this kind of stuff). The strength of a bone depends on the cross-sectional area of the bone, which increases as the square of the radius of the animal. But the weight of the animal increases as the cube of the radius. So the weight of the creature increases faster than the strength of its bones. A very big animal can't carry its own weight (so a blue whale can survive only in the ocean, where the water carries its weight). The long leg bones of an elephant could possibly break in a 1-metre vertical fall, if it landed squarely on its legs.

So, weight for weight, the smallish bones of a cat are much stronger than our bigger bones. That's the second factor helping the cat survive a fall.

Survival 3: Body Alignment

But to answer the big question, *"Why is it safer for a cat to fall from a really tall building than from a shorter building?"*, we have to look at a third factor — the Body Alignment of the cat.

We are now in the Land of Theory. After all, nobody has filmed the entire flight of a cat falling 32 storeys with a high-speed camera (don't do this experiment at home, folks).

We know that cats reach their terminal velocity after five or so storeys.

My personal theory is that once a sensible falling cat has reached its terminal velocity, it will realise that it is not accelerating any more, and will put its energies into aligning its body.

Suppose a falling cat gets its legs pointing straight down and its body parallel to the ground. On hitting the ground, the arms and legs absorb some of the energy of the impact — they bend, but don't break. A bit more energy will be absorbed by stretching the muscles, ligaments and tendons in each leg. A squillionth of a second later, the rib cage hits, and the rest of the energy is spent

THE PHYSICS OF A SELF-RIGHTING CAT

If you hold a cat with its paws upwards, and then drop it, the cat quickly twists around in mid-air so that it lands paws downwards (don't do this experiment at home, folks). How can a cat do this if it doesn't have anything "solid" to grab onto?

The answer lies in Conservation of Momentum. When you travel in a straight line, you just talk about Momentum. But when you travel in a circle, you talk about Angular Momentum.

If a cat twists its front legs 30° to the right, then, to balance Angular Momentum, the back legs will twist 30° to the left, providing that the legs are the same length. The special trick of the self-righting cat is that it can quickly stretch out, or pull in, the legs that it happens to be twisting. If a pair of legs is stretched out, the cat can rotate them 30°, but if they are pulled in, it can rotate them 60°. (Ice-skaters use this technique to speed up during a spin.)

Let's follow a falling cat.

It begins its fall on its back with both front and rear legs pointing straight up, and stretched out. Let's call this starting position 0°.

First, the falling cat pulls in its front legs, and rotates them 60° clockwise (CW). To conserve momentum, the rear legs then have to rotate anti-clockwise (ACW). If the rear legs were pulled in, they would rotate 60° ACW. But the cat cleverly keeps its rear legs stretched out, and so they rotate only 30° ACW. The plunging pussy now has the front legs 60° CW, and the rear legs 30° ACW — and all the angular momentum equations are balanced.

Now for the second stage.

The cat stretches out its front legs, and rotates them 30° ACW, taking them to 30° CW. The rear legs have to rotate CW to balance the angular momentum. The cat pulls in the rear legs, which then rotate 60° CW back to 30° CW.

The falling cat now has both front and rear legs at 30° CW. It is still falling towards the ground, but has managed to rotate its whole body 30° CW.

If it repeats the above cycle another five times, it will rotate its legs a total of 180°, so that they are pointing towards the ground. The cat is amazingly agile and quick, to be able to do all this in a metre or so of free-falling.

the physics of a self-righting cat

puss falls out of window on back with both front and back legs pointing straight up & stretched out

plunging pussy

meeooooooWWWW

front legs back legs

1. fall begins on back with legs up & stretched out

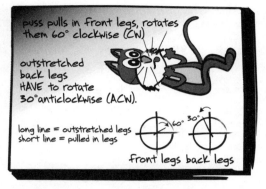

puss pulls in front legs, rotates them 60° clockwise (CW)

outstretched back legs HAVE to rotate 30°anticlockwise (ACW).

long line = outstretched legs
short line = pulled in legs

60° 30°

front legs back legs

2. the rotation begins...

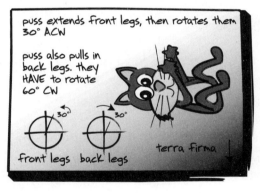

puss extends front legs, then rotates them 30° ACW

puss also pulls in back legs. they HAVE to rotate 60° CW

30°   30°

terra firma

front legs back legs

3. ...puss' whole body rotates 30° CW!

puss has rotated its body 30°. after he repeats the process 5 more times, the legs point toward the ground.

still plunging pussy

30°

puss body rotation meter (pussometer™)

4. repeat cycle 5 times...

bending, but not breaking, the ribs. The cat will spread the impact over its whole body, not just on its legs. The cat will land purrfectly.

The one cat that fell 32 storeys onto the concrete pavement had only minor injuries — a slightly collapsed lung and a chipped tooth.

Useful Advice

But this raises three issues. First, be careful when you next visit New York — it's raining cats at one per day. Second, if you fall out of a tall building, make sure you take a cat with you to show you what to do. And third, maybe the cats didn't jump after all — maybe they were pussed.

HUMAN FALLS IN THE UNITED STATES

Back in 1986, Warner and Demling estimated (in a paper in the *Annals of Emergency Medicine*) that free falls killed about 13 000 Americans each year. About 50% were accidental, 20% were suicide attempts, another 20% related to crimes, and the causes of the remaining 10% were not known. High blood levels of alcohol (over 0.10%) were involved in one third of all cases of falling adults.

Warner and Demling found that it was better to slow down over a greater distance. The G-forces on your internal organs would not be so high, and your chances of surviving were greater. There was a case of a female parachutist surviving a 2800-metre (1.74-mile) fall when her parachute failed to open. She was lucky enough to land on soft ground, and, according to the fireman, "*put a good 12-inch (30-cm) hole into the ground*". But, in general, any fall greater than six storeys onto solid ground will kill an adult.

Children tend to survive these falls better. There are a few reasons for this. First, their terminal velocity is not as high as that of adults. Second, their bones are, weight for weight, stronger. Third, their bones are not fully calcified, and are more likely to bend than break.

The injuries a person will suffer depend on the height of the fall, and how they land. So 80% of buttocks landings lead to pelvic fractures, 60% of head-first landings cause skull fractures, while 25% of feet-first landings produce fractures of the ankles or feet.

Humans take about 150 metres (500 ft) to reach their terminal velocity of 200 kph (120 mph). The famous Mexican Acapulco cliff divers, leaping from 41 metres (135 ft) reach a top speed of just 105 kph (65 mph). They are trained to hold their arms stiffly over their heads, to absorb much of the entry shock. (But the amateur human falling onto water will do better landing feet-first.)

Over 720 people have died after jumping off the Golden Gate Bridge in San Francisco. At the midpoint, the bridge is 76 metres (250 ft) above the water. This gives an impact velocity of 120 kph (75 mph). All of the known survivors hit the water feet-first.

HUMAN FALLS FROM THE SYDNEY HARBOUR BRIDGE

Between March 1930 and October 1982, 92 people fell from the Sydney Harbour Bridge. They fell some 59 metres (195 ft) to the water below. They hit the water at an estimated 110 kph (68 mph).

Only 14 survived — a survival rate of about 15%. Landing feet-first improved one's chances of living. Half of the survivors had lung damage.

ENERGY TO INJURY

It's not the falling that hurts you, it's the stopping. The more that you can soften and stretch out that sudden "stopping", the lower your G-forces and the lesser your injuries.

Suppose that you fall just one storey onto hard concrete. The initial impact is the first collision. Your brain keeps moving inside the skull, and then stops over a distance of 6 mm ($\frac{1}{4}$ in.). This is the second collision. Your 1.5-kg (3.3-lb) brain experiences a deceleration of 720 G, so it suddenly weighs 1080 kg (2380 lb).

The structure of the brain (and other organs, for that matter) is not strong enough to support this sudden "weight". It distorts, and various internal structures (such as arteries and veins) can "break". Blood can leak out, causing damage to the brain. These enormous G-forces can also make individual cells (which are little bags of fluid) burst open and die.

But suppose that you fall three storeys onto mud, and stop over a distance of 20 cm (8 in.). Your body experiences a deceleration of "just" 67 G, and you have a much better chance of surviving.

But the lungs can be damaged at fairly low G-forces. The ribs and the outside of the lungs are very slippery, so they can slide over each other easily. But there is a "potential space", called the "pleural cavity", between the ribs and the lungs. Air can leak into this cavity from the lungs if the delicate lung tissue is torn. This external air stops the lungs from opening up fully, and partially collapses them, so less air gets in. This condition is called pneumothorax (literally, "air-chest"). In a very severe pneumothorax, you can't get enough air to stay alive.

The pneumothorax is often easy to treat. You just suck the air out with a syringe. This is called thoracocentesis (literally, "chest puncture"). If the injury is not too severe, your body will repair the leak. In a severe pneumothorax, surgery is necessary to repair the damage.

CATS LOVE CATNIP

Catnip is a herb related to garden mint. It grows in ditches or borders, is about 50 cm (20 in) high, and has white flowers that are spotted with crimson.

Cats love catnip, especially when the plant is damaged. They will thrust their face down to the ground, and sweep it from side to side. They also prick up their ears, and push their front legs forward, while extending their claws.

The chemical that causes this behaviour is called nepalectone. It was discovered in oil of catnip in the 1940s. Nobody knows why nepalectone makes most cats so happy.

Edgar Booth and Phyllis M. Nicol, *Fundamental Laws and Principles with Problems and Worked Solutions* (Tenth Edition), 1946, Australasian Medical Publishing Company Limited, pp 366–367.

Jeremy Cherfas, "How to Thrill Your Cat This Christmas", *New Scientist*, No. 1592–93, 23–31 December 1987, pp 42–45.

Jared M. Diamond, "Why Cats Have Nine Lives", *Nature*, Vol. 332, 14 April 1988, pp 586–587.

Gordon W. Robinson, "The High Rise Trauma Syndrome in Cats", *Feline Practice*, Vol. 6, No. 5, September 1976, pp 40–43.

"Terminal Velocity", *Discover*, August 1988, p 10.

KG Warner and RH Demling, "The Pathophysiology of Free-Fall Injury", *Annals of Emergency Medicine*, Vol. 15, 9 September 1986, pp 1088–1093.

W. O. Whitney, and C. J. Mehlhaff, "High-Rise Syndrome in Cats", *Journal of the American Veterinary Medical Association*, Vol. 191, No. 11, 1 December 1987, pp 1399–1403.

REFERENCES

REAL WHEEL ANIMALS

A nimals can squirm and swim, jump and hop, and can even flip, flop and fly. And now we've found a few animals that make like a wheel — and they're real!

The Wheel

The wheel is one of the greatest inventions of the human race.

The wheel is really handy if your civilisation has lots of flat, wide open spaces, and strong draft animals to pull a wheeled cart. Classical Rome had a magnificent network of 80 000 km of paved roads, and they had all kinds of wheeled vehicles. However, the wheel is not much use in soft sand or in hilly terrain. The early Egyptians and the peoples of South America mostly used river transport and pack animals. The South Americans used the wheel in children's toys, but never for carrying loads.

Wheel Invented Once

We're not absolutely sure, but it seems that the wheel was invented just once, and then the knowledge of the wheel spread across the world. It does not seem as though many different peoples each invented the wheel for themselves.

Wheels seem to have started off in the bygone civilisation of Sumeria, in the fertile area between the Euphrates and Tigris Rivers — what is now modern Iraq.

The Potter's Wheel

The very first wheel that we know of is the potter's wheel, dating from about 3250 BC. It was used to make clay into round shapes, like dishes or vases, and it's still used today.

The next piece of evidence is a clay tablet, dating from around 3150 BC, with a picture of a wheeled cart on it. It was found in the courtyard of the Eanna Temple in the city of Uruk in ancient Mesopotamia. This was the first wheel thought to carry a load. Many archaeologists say that the wheel had probably already been around for a

EARLIEST PAVED ROAD

As far as we know, the Egyptians didn't use wheeled vehicles pulled by draft animals until around 100 BC. They mostly used water transport.

Before 100 BC, the Egyptians still had to get their heavy loads to the water barge. How did they carry them across the sandy deserts?

In 1994, Dr James Harrell (a Professor of Geology at the University of Toledo in Ohio) and Dr Thomas Bown (a research geologist) reported that they had discovered the world's earliest known paved road, in Egypt.

This road was built around 4600 years ago. It is 2 metres wide, and runs across the desert some 68 km southwest of modern Cairo. It leads from a large basalt quarry to a lake, which in ancient times led into the Nile. This road is made of slabs of sandstone and limestone, and even some logs of petrified wood. Apart from some construction ramps that were used to make the Pyramids, there are no other paved roads known in ancient Egypt.

thousand years by then, but we don't have any definite proof of that . . . yet.

The First Cart Wheel

That tablet just had a picture of a wheeled cart. But in the ancient city-state of Kish, we have found actual physical remains of chariots with solid wheels, dating back some 4650 years. The chariot wheels were about 0.5–1 metre (1.6–3.3 ft) across. These solid wheels were cut from three planks of wood that had been clamped together with wooden struts. They had copper nails to hold the leather tyres on.

While a solid wheel could carry a load, it couldn't move very quickly. If you wanted organised wheeled warfare, you needed a lighter, faster wheel. One way to reduce the weight of the wheel was to carve sections out of it. This led to the radial-spoked wheel around 2000 BC.

The spoked wheel appeared in either Northern Syria or Anatolia, and was probably invented by the Hittites or the Mitanni. Sometimes the spokes were made from bronze, to make the wheels stronger and lighter, so the chariot could go faster. Unfortunately, this meant you needed better roads. So 4000 years ago, Assyrian road builders used bronze pickaxes to cut away the side of a steep hill to build one of these better roads.

We also know that the Hyksos people used spoked wheels on their horse-drawn war chariots to invade Egypt in 1680 BC.

Wheel Reaches Asia and Europe

According to ancient Indian folklore, barbarians, who were skilled in using metal, arrived in India with horse-drawn chariots around the 14th century BC. The Indians seized upon this new idea of a wheeled vehicle, and adapted them into everything from fast, light, two-wheel war chariots to slower, four-wheel coaches and carriages.

In China, the light chariot, used for hunting, appeared around 1200 BC. It had the new high-tech spoked-wheel, with between 18 and 26 spokes per wheel.

The saying goes, "*All roads lead to Rome*", and that's because all of the Romans' roads led out of Rome. To use their magnificent road network, the Romans had a huge variety of wheeled vehicles — slow, heavy freight wagons, passenger coaches, light-weight chariots for racing, warmongering and hunting, fashionable gigs for going out on the weekends, covered carriages, and two-wheeled farm carts. In fact, the only real improvements in the wheel since the time of the Romans are the invention of ball bearings and rubber tyres.

Efficient Wheels

We humans use wheels because they can make us go faster or lighten our workload.

In a typical marathon, a wheelchair athlete can beat the fastest runner by about 20–30 minutes.

A "human-being-riding-a-bike" is the most efficient way to travel, measured by the amount of energy needed per kilometre of travel and per kilogram of vehicle weight. A person on a pushbike is five times more efficient than a car or passenger jet, 15 times more efficient than a running dog, and 400 times more efficient than a cockroach.

If wheels are so useful, then why didn't Mother Nature make more use of them in animals?

Disadvantages of the Wheel

It turns out that wheels have a few dis-advantages. First, wheels roll easily on hard surfaces, but they slow down enormously on thick grass, soft sand, or fluffy carpet.

Second, wheels run into problems with vertical obstructions. Try pushing your supermarket trolley over a gutter or curb.

WHEELED SPACE VEHICLES

The first wheeled vehicle used outside of the Earth was the unmanned Lunokhod I. This Soviet vehicle began its travels on 17 November 1970, and covered some 10.53 km (6.54 miles) on the Moon, until it broke down on 4 October 1971. It was a slow-moving vehicle, because it was controlled from Earth.

Both the speed and distance records on the Moon were set by John Young from the Apollo 16 Mission. He drove the Lunar Rover at a maximum speed of 18 kph (11.2 mph) downhill. Overall, he covered a total distance of 36 km (22.4 miles).

In general, the highest obstruction any vehicle can climb is about half the radius of the wheel. However, if your vehicle has a flexible frame and can shift its centre of gravity, you might be able to climb over an obstruction as high as the radius of your wheel.

The third problem with wheels is that they're not very good at making tight turns. This is really obvious when you follow a large truck or bus along a narrow road with hairpin bends.

And when you think about a typical day's travel, you'll find that most of the time we humans use legs to get around.

Animals that use "rotation" to get around are very rare. There are bacteria that swim around with a microscopic spinning propeller called a flagellum, but recently we discovered animals that spin like a wheel.

Animals with Roller Skates?

I don't mean animals that have little wheels on them — like little retractable roller skates. It would probably be too hard to have wheels with nerves and blood vessels running to them. They would get all tangled up as the wheel turned. But we humans do have dead parts without nerves or blood vessels, like our hair and fingernails. It would be difficult, but definitely possible, to build wheels out of these dead parts.

However, we have never found any living animals, or fossils of extinct animals, that have wheels on them.

What we have found are two animals that will roll themselves along — one on the seashore, and the other on dry land. Each of these animals actively uses its muscles to roll like a wheel. They don't just roll down a hill, and they don't just exist in legends.

Animals Roll Down Hills

So they're quite different from animals like the little 10-cm (4-in) Web-Toed Salamander (*Hydromantes platycephalus*), which lives on the steep sides of mountains in the Sierra Nevada of California. It will coil itself up into a little ball when it's disturbed or startled. If it happens to be on a slope, it will roll downhill. In fact, the web-toed salamander will tuck its legs in, and wrap its tail around its body, to make itself even better at rolling. This salamander uses gravity, not muscles.

Stephen Deban, a herpetologist (reptile scientist) from the University of California at Berkeley, tested 16 species of salamander. Some of these salamanders, when they were startled, would coil into a ball, but none of them would allow themselves to roll downhill. In fact, they would stretch out their legs to stop themselves.

There are also some desert spiders that use gravity to escape. These spiders fold up their legs and roll downhill to get away from danger.

The Pangolin of Southeast Asia gets its name from the Malayan words meaning "rolling over". It's an armoured mammal, also called the Scaly Anteater. The pangolin is another animal that, if under threat, will roll itself into a ball, and if it's on a slope, roll down the hill. There are eight species, located in tropical Asia and Africa. They range from 30–90 cm (1–3 ft) long, and weigh between 5–27 kg (11–60 lb). They mostly eat termites, but will eat ants and other insects as well. The pangolins are delicious to humans as well as other animals and they have no real defence against their predators. All they can do is emit noxious odours from their anal glands and roll themselves up into a little armoured ball.

Mythical Rolling Animals

The recently discovered rolling animals do not fit the description of mythical rolling animals. The "Hoop Snake" of the American Midwest is supposed to put its tail into its mouth, joining itself into a complete circle. It rolls along the countryside and can suddenly straighten out to form a javelin. It then hurls itself, fangs first, into the victim.

The Dutch artist Escher dreamed up another mythical animal he called the "Curl Up" creature. This animal would walk using its three legs on each side, but when it was in a hurry, it would curl up into a wheel and roll along.

Let's leave the animals that will roll down a hill, and the mythical rolling animals, and talk about real animals that use their muscles to roll away from danger.

Real Wheel Watery Animal

Roy L. Caldwell, an animal behaviourist at the University of California at Berkeley, discovered the first rolling animal in 1979. It's a tiny ocean crustacean called *Nannosquilla decemspinosa*. It's about 2–3 cm (1 in.) long and lives in holes in the sandy beaches on the Pacific side of Panama. Its specialty is the backward somersault. This crustacean is related to the prawn that you chuck on your BBQ and, like other crustaceans, it has a hard outer shell and legs with joints, but it also has gills on appendages that are attached to its abdomen.

Dr Caldwell put one of these little crustaceans under his microscope — but when he looked through the lens, it was gone! So he put another one under the

real wheel watery animal

1. introducing...nannosquilla decemspinosa

2. far, far away from home

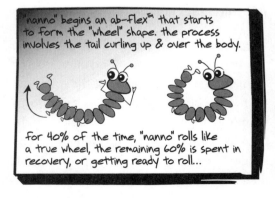

3. the rotation begins...

4. rollin' home to my baby!

microscope and, by the time he looked through the lens, it too was gone. But as he looked up, Caldwell saw the tiny creature with his naked eyes rolling down the laboratory bench.

Nannosquilla lives in long skinny holes that it burrows into the shifting sands. There are two advantages to long skinny holes. First, they are stronger than long fat holes. Second, it lines its hole with mucus, and a skinny hole needs less mucus than a fat hole. So our little rolling animal has evolved to become long and skinny.

Because it's so long and skinny, it's virtually paralysed when it's on the sand. It does have three pairs of legs at the front, but they're so tiny and weak that they're pretty much useless. So when the waves throw them up on the beach away from their burrow, they curl their head to their tail and travel along the sand for a distance of up to 2 metres (6 ft). They can do a maximum of 40 backward somersaults, spinning at 72 revolutions per minute. Their ground speed varies during each cycle, but averages out to 0.13 kph (0.08 mph).

Now when Caldwell and his team in the Department of Integrative Biology actually videotaped the small wheeling animal, they found that for 40% of each revolution, *Nannosquilla* actually rolled like a true wheel! The remaining 60% of each revolution was spent recovering from a roll, or getting ready to roll. When it came out of a roll, it would lie down on its back, flat on the sand. It would leave its head still, lift up its tail and curl it over until it touched its head. Then it would start rolling like a wheel again.

Real Wheel Land Animal

The only known land creature that will deliberately roll itself away from danger is the caterpillar of the Mother-of-Pearl Moth, *Pleurotya ruralis*. This research was done by John Brackenbury at the University of Cambridge in the United Kingdom.

Most caterpillars have some sort of defence against attackers, such as a warning set of colours, irritant hairs, or tasting really bad. The mother-of-pearl caterpillar doesn't have any of these defences, and it is really slow at walking. So, in extreme danger, it relies on rolling away at 40 times its walking speed.

Regular Caterpillar Walking

These caterpillars have a body made up of 13 segments. They have legs attached to each of the end segments (No. 13 and No. 1), as well as four sets of legs in the middle on segments six to nine. When they're walking forward, you can see the characteristic travelling "hump" or "wave" move along the caterpillar's back. As the wave comes along, it lifts a segment up from the ground, squashes it into the segment in front of it, and then lowers it back down to the ground — slightly forward of its previous position. In each cycle of taking one step forward, each foot is on the ground 65% of the time. This does make the caterpillar very stable, but it also stops it from walking rapidly. This is why caterpillars walk so slowly: approximately 1 cm per second, or one tenth of the speed of other insects of the same weight. That's about 0.036 kph (0.022 mph).

real wheel land animal

the characteristic "wave" of the caterpillar walking occurs when it lifts segments up & squashes them forwards & then lowers it slightly forward of its previous position.

movin' forwards

in each cycle of taking one step, each foot is on the ground 65% of the time.

1. the basic caterpillar

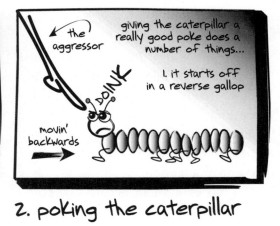

the aggressor

giving the caterpillar a really good poke does a number of things...

1. it starts off in a reverse gallop

DOINK

movin' backwards

2. poking the caterpillar

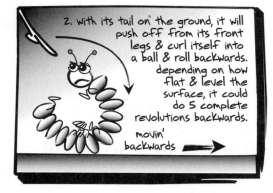

2. With its tail on the ground, it will push off from its front legs & curl itself into a ball & roll backwards. depending on how flat & level the surface, it could do 5 complete revolutions backwards.

movin' backwards

3. the defensive roll...

But if you provoke this caterpillar by poking it on the head or chest, things get interesting . . .

Emergency Caterpillar Walking

If you give the mother-of-pearl caterpillar a little poke, it will walk backwards. This backward-walking looks exactly the same as the forward-walking, except, obviously, in the opposite direction.

If you give the caterpillar a medium poke, the wave will move across its body much more quickly, the legs will spend more time in the air, and it will actually retreat from you at a fast gallop.

PRAYER WHEEL

The Tibetan Buddhists found a unique use for the wheel.

The Prayer Wheel is a hollow, embossed metal cylinder. Inside there is a consecrated written prayer, or mantra. Turning the wheel once by hand is considered the same as saying the whole prayer aloud once.

But if you give it a really good poke, it will start off in its reverse gallop. Then, with its tail on the ground, it will push off from its front legs, curl itself into a ball, and roll backwards. Depending on how flat and level the ground is, a decent push will set it off into five complete revolutions, travelling at about 40 cm per second, or about 1.44 kph (0.9 mph).

Now maybe there's a lesson in this for us humans. Engineers have been working with wheels for thousands of years. Over the last 20 years or so they've also been working on robots with many legs.

Maybe we can take a lesson from Mother Nature and design the ultimate wheeled vehicle that is so advanced it has only one wheel!

REFERENCES

John Brackenbury, "Caterpillar Kinematics", *Nature*, Vol. 390, 4 December 1997, p 453.
Jared Diamond, "Why Animals Run on Legs, Not on Wheels", *Discover*, September 1983, pp 64–67.
Robert Full et al., "Locomotion Like a Wheel?", *Nature*, Vol. 365, 7 October 1993, p 495.
Robert G. Rogers, "How Wheels Could Walk", *Discover*, October 1983, p 77.
"Salamanders That Roll", *Discover*, August 1995, p 21.
Carl Zimmer, "See How They Run", *Discover*, September 1994, pp 64–73.

MURPHY'S LAW ON EARTH

Everybody has heard of Murphy's Law. It gets the blame whenever anything goes wrong — like toast falling buttered side down, or every other queue at the supermarket moving faster than yours.

Murphy's Law Isn't . . .

But most people, when they're referring to Murphy's Law, are actually thinking of the rather pessimistic Finagle's Law, which is *"Anything that can go wrong, will go wrong"*. Finagle's Law basically says that there is no point in trying, so you may as well give up right now.

But Murphy's Law is very different, and is actually a message of hope. Not only does it warn you what things can go wrong, it even tells you how you can work around them so as to save yourself from disaster.

It All Began after World War II

The aerospace industry, where planes and spacecraft can move very rapidly under very harsh and extreme conditions, is an unforgiving environment. And that's where Murphy's Law was first recognised.

After World War II, various air forces started using jet engines, which were much more powerful than the old piston engines. Suddenly, pilots were reporting difficulty staying conscious, because the G-forces were draining blood from the brain.

G-Forces

It has become a convention to measure acceleration or deceleration in "G"s. On the surface of the Earth, we all experience the normal gravitational force of 1 G acting on our body. We can briefly experience 1.5 G when a commercial jet takes off, or when a lift suddenly rises. At 1.5 G, we "weigh" 1.5 times our normal weight.

Positive G will push you into the seat — the pilots call this "eyeballs in". Negative G lifts you out of the seat — they call that "eyeballs out". It's slightly more comfortable to have your eyeballs pushed in, and so pilots are usually better at resisting high positive G-forces than negative G-forces.

When pilots want to turn the plane to the right, they can't simply turn the nose to the right. The wing on the outside of the curve would travel slightly faster than the wing on the inside of the curve, therefore generating more lift. This creates too many problems. Instead the pilot "banks" the plane — the right wing drops and the left wing rises. But that creates an artificial gravity force that pushes blood from their brain and abdomen down into their feet.

What G-Forces Do to You

The average fit person begins to feel their limbs getting heavy at around 2–3 G (when they weigh two to three times their "real" weight). It's also hard to keep your head up at 2–3 G. In the average Space Shuttle launch, the crews experience about 2.5 G.

Around 4 G, white specks appear in your vision. The retina is very oxygen-hungry, and it's not getting enough blood to feed it the oxygen it needs. Your colour vision turns into black-and-white — a phenomenon the pilots call "greyout". As if that wasn't bad enough, you also lose your peripheral vision, leaving only a small circle of vision dead ahead. At this G-force, it's impossible to lift your head.

As the G-forces slowly increase, you lose all vision. Even though you remain conscious and are able to listen and reply, you can't see. At 8 G, if you are still conscious, your arms and legs are so heavy that you can't lift them.

G-LOC Can Kill

In general, 5 G for five seconds will make you unconscious. It's called "G-induced Loss of Consciousness" or G-LOC. Modern jet fighters can make a sharp turn very quickly, and generate very high G. When these planes suddenly snap into a sharp turn, the pilot doesn't slowly progress from feeling heavy, to a greyout, then to G-LOC. Instead, without any warning, the pilot can instantly lose consciousness. The United States Air Force believes that in the last 10 years alone, 20 fatal crashes have been caused by unexpected G-LOCs.

In centrifuge tests, the effects of G-LOC are bizarre, and very dangerous. Sometimes, the pilots' arms and legs jump around as if they are having a major epileptic attack. Even after the pilots have come out of G-LOC, they are completely incapacitated and unable to move for 15 seconds. For the next 15 seconds, they can move their arms and legs, but they are so confused and disoriented that they can't do anything sensible. So, for about 30 seconds after G-LOC, nobody is flying the plane.

Fighter Pilots Need G-Suits

Early aircraft would collapse or disintegrate around the 2–2.5 G mark. But modern military aircraft can take up to 9.5 G positive and 5.5 G negative. The new generation Eurofighter should even be able to inflict 12 G on the pilot.

But unprotected pilots can't withstand these high G-forces. Fighter pilots now wear special G-suits, which force the blood back into their upper body, especially to the brain. The typical G-suit has five inflatable

THE G-SUIT

The G-Suit was invented by Professor Frank Cotton, a slightly eccentric physiologist from the University of Sydney.

In the early stages of World War II, in 1940, pilots from both sides were flying faster and pulling harder turns. As the blood left their heads, they'd suffer from temporary blindness, and sometimes unconsciousness, which, of course, could get them killed.

The German Luftwaffe had known about this for some time. In the mid-1930s they had built the first centrifuge that could take a human being. But even though they were able to apply enough G-forces to make their pilots unconscious, they couldn't come up with a solution to keep them conscious.

Late in 1940, Professor Cotton was a Reader in Physiology at the University of Sydney. He happened to read an article in his evening newspaper saying that blackouts in pilots were a *"crucial element"* in aerial combat. He wrote in his notes *"Inside one minute I could see the solution. It was simply a matter of combining applied dynamics . . . and physiology."* All he needed was a suit with a set of inflatable airbags, to squash the pilot's legs and abdomen, and push the blood back to the brain.

His first G-suit was made from two women's rubber bathing suits. It exerted high pressure around the lower legs, and lower pressure around the abdomen. The next version, the Mark I G-suit, gave pilots the ability to withstand G-forces 30% higher than if they were not protected.

Professor Cotton met with American officials in Washington, D.C., in December 1941. By this time, Canada already had its own G-suit, but it used water instead of air, so it was heavier and more cumbersome. The Americans adapted Professor Cotton's design. With his assistance, by 1944, United States Navy pilots were wearing gradient-pressure pneumatic G-suits.

In 1942, he said, with an unerring sense of the future, *"As a result of my work, man's now going to the moon!"*

bladders — one across the stomach, one on each thigh, and one on each calf. When these bladders inflate, they push the blood back upwards.

The G-suits can keep a pilot conscious up to about 10 G. But that knowledge was hard-won — and accidentally gave us Murphy's Law.

Rocket Sled Tests

Soon after World War II, the United States Air Force ran a test series (Project MX981) to see what sort of G-forces a human being could withstand. They used volunteers strapped into a rocket sled at Muroc (now called Edwards Air Force Base) in

California. The attraction of Muroc was that it had a 610-metre (2000-ft) standard gauge railroad track supported on a heavy concrete bed. This track had been originally used for research into V-1 rockets. The rocket sled didn't use wheels, because they might disintegrate at high speed. Instead, it had four "slippers" that slid along the track. The sled would accelerate up to speed and then stop suddenly. It had 45 brakes, giving it probably one of the most powerful braking systems ever built. It could provide controlled decelerations at almost any desired rate.

The first rocket sled decelerator run took place on 30 April 1947, using ballast instead of a person. That was lucky because the sled ran off the tracks.

Humans on Rockets

The first rocket sled run with a human was in December 1947. In the first 16 rocket sled runs, the volunteer was always facing backwards. Full instrumentation and recording of data began in August 1948. Experimenters also tried some forward-facing runs at this time. The goal was to collect enough data to work out what kind of harness or seat belt would give the greatest protection to a pilot.

One of the most frequent volunteer human torpedoes was Major John Paul Stapp, who was also a medical doctor.

Murphy's Job

Our hero, Air Force Captain Edward A. Murphy Jr., had designed a harness that strapped onto the volunteer. This harness held 16 sensors to measure the acceleration, or the G-forces, on different parts of the brave volunteer. As luck would have it, there were two ways that each sensor could be installed.

On one occasion, the rocket sled took off and stopped suddenly, generating 40 G. Under 1 G, the average person weighs about 70 kg, but under 40 G, they weigh 40 times more — about 2.8 tonnes (2.75 tons). The "pressure" of 40 G gives an enormous

HUMAN G-FORCES

When you take off in a jet, all you're feeling is 1.5 G.

But the ejection systems in jet fighters can generate accelerations, for about 0.1 second, of greater than 30 G. In the early days when explosives were used, the G-forces were higher, but now rockets give lower G-forces. The official measured record for a human experiencing a high G-force and surviving is 86 G in a rocket sled. Your "average" 70 kg (154 lb) male would suddenly "weigh" 6.0 tonnes (5.9 tons).

amount of deceleration — enough to push your ears onto the front of your head.

The unfortunate Major Stapp had bloodshot eyes, and was bleeding from a number of bodily orifices after his ride. He was barely able to speak, but he had to know, so he asked, *"How many G did the sensors read?"* The unhappy technician replied, *"Zero."*

Major Stapp had been strained in vain.

Murphy's Law Happens

He called for Captain Murphy, who immediately flew in from Ohio. When Captain Murphy examined the sled, he found every single one of those 16 sensors had been installed the wrong way round.

Murphy's Law Proclaimed

In a voice like thunder, Edward A. Murphy Jr. proclaimed, *"If there are two or more ways to do something and one of those results in a catastrophe, then someone will do it that way."* This is the True and Original version of Murphy's Law.

It's optimistic because of that little word, *"If"* in *"If there are two or more ways . . . "* To avoid a catastrophe, all you have to do is make sure that there is only **one** way to do something, and that the one way is correct.

So once he had realised the significance of Murphy's Law, Murphy himself redesigned the G-force sensors so that they could be installed only one way — and that particular problem was solved forever.

Murphy's Law Publicised

George Nichols, the Project manager, realised that he had just heard a Great Thing, and immediately gave it the name "Murphy's Law". The next day Murphy's Law was unveiled at a press conference about the

HOW TO USE MURPHY'S LAW

If you design something that can be installed only one way, then it can't be put in the wrong way.

Consider the electrical power in your house. You can tap into it with a three-pin electrical power plug. The plug is not symmetrical in all directions. There's only one way to insert it into an electrical power socket.

But sometimes, it doesn't matter which way a thing goes in, so long as it goes in. So another way around Murphy's Law is to design something so that it doesn't matter which way it goes in.

Consider the key. Your average *car* key has equal bumps on both sides, so it doesn't matter which way you put it in — it always works. But your *house* key has bumps only on one side. This means that half the time, you'll put it in the wrong way.

Could it be an example of Murphy's Law that over half a century after the law was first proclaimed, house key manufacturers don't use it?

MAKE YOUR CAR WEAKER

Crash testing a real car is expensive. A lot of preliminary design can be done with a Crash Simulation Program. The three most popular programs used today are PAM-CRASH (designed in Europe, and used in Europe and Japan), LS-DYNA3D (popular in the United States) and RADIOSS (developed in France). They are all based on military applications developed in the United States in the late 1960s.

In 1995, a BMW design team used PAM-CRASH to improve the side-impact safety of the 5-series BMW.

The roof of a four-door sedan is joined to the floor by six pillars — the A-pillars (each side of the windscreen), the B-pillars (between the front and back doors) and the C-pillars (behind the back doors). Physical crash tests showed that when the car was hit from the side, a small section of the B-pillar close to the floor would fail.

The answer seemed clear: simply make this section stronger — no need to test the results with another expensive crash test. But one of the engineers insisted on seeing what the PAM-CRASH computer simulation would show.

The team was surprised to see that making the B-pillar stronger would make the car less safe in a side impact. PAM-CRASH showed that the B-pillar would now collapse higher up, allowing side-penetration of the passenger compartment closer to the vulnerable abdomen, chest and head of the passengers.

By 1996, the design team had done 91 PAM-CRASH simulations and two physical crash tests, and the side impact safety was now 30% higher. The two physical crash tests cost $US300 000, which was more expensive than the 91 virtual crash tests.

It goes to prove that old saying in physics, "*For every problem, there is a solution that is simple, clear, easy to understand, intuitively correct . . . and 100% wrong!*"

rocket sled test. Major Stapp claimed that part of the reason that Project MX981 had such a good safety record (i.e., nobody died) was that all the team believed in, and followed, Murphy's Law.

A few months later Murphy's Law began to be mentioned in aerospace manufacturers' advertisements, and even by the Flight Safety Foundation.

As Murphy's Law spread across the planet, two things happened: people forgot that there was a real person called Murphy, and the Law got modified into the pessimistic version of "*If anything can go wrong, it will.*"

The fact that many people confuse the optimistic Murphy's Law for the pessimistic Finagle's Law is proof that Murphy's Law can even act upon itself.

THE FAMOUS DR STAPP

The work of Dr Stapp has saved hundreds of thousands of lives. The seat belts in your car follow directly on from his research.

Stapp proved that a human could withstand much higher G-forces in a backward-facing seat than in a forward-facing seat. As a direct result, all United States Air Force Military Air Transport planes were fitted with backward-facing seats.

Dr Stapp also showed that pilots could survive massive decelerations if they wore the appropriate harness, and if the seat did not tear loose. As a result, the mounting requirement for fighter seats was increased to 32 G.

He also developed the "side saddle", or sideways-facing, triangular-shaped harness for fully equipped paratroopers sitting side by side in Air Force planes. It was made of nylon mesh webbing, had a shoulder strap, and replaced the simple lap belt.

Stapp took more rocket sled rides than any other person. He showed that a person could withstand at least 45 G in the forward position, if they wore the appropriate seat belts. This is thought to be the highest sustained G-force voluntarily undertaken by a human.

He also developed an addition to the lap-sash belt — another two straps, going down from the centre of the lap belt, to hold the thighs in position. This spread the load over the stronger parts of the body — the shoulders, hips and thighs, instead of the belly.

Dr Stapp also put himself through wind-blast experiments to see if a pilot should remain with the plane if the canopy comes off. So he flew in a plane at 917 kph (570 mph), without the canopy. He survived the wind blasts with no injuries.

On 10 December 1954, now promoted to Colonel Stapp, he achieved the land speed record in a rocket sled (the Sonic Wind 1) at Holloman Air Force Base, New Mexico. He reached a speed of 1017 kph (632 mph).

He paid the price for all these achievements, though — with his body. He suffered broken ribs, several haemorrhages of the retina, and two broken wrists.

Stapp is now in his 80s, and still gives a lecture at the Stapp Car Crash Conference, named in his honour and held each year since 1955. He is truly a remarkable person.

G-FORCES IN CARS

The fronts of modern cars are designed to crumple, so that the passengers will be protected in their "safety cage" passenger compartment. This is the old "cardboard box" versus "army tank" problem.

If the car is only as stiff as a cardboard box, it will instantly collapse and absorb very little of the impact forces, which all get passed on to the passengers. If the car is as stiff as an army tank, it will not collapse at all. Again, it will absorb very little of the impact forces, passing them on to the passengers. The car manufacturer Peugeot has pointed out that a head-on collision between two tanks is likely to be fatal to the occupants at 20 kph (12 mph).

So your car needs a "stiffness" somewhere between a cardboard box and a tank.

This involves a lot of careful juggling in the design of your vehicle. Your average car has a relatively strong frame, relatively weak body panels, and a few large lumps of metal like the engine and gearbox.

Under the Australian New Car Assessment Program (NCAP), the typical G-forces in a 56 kph (35 mph) full-frontal collision into a fixed concrete barrier reach a peak of 30–40 G, when measured at the B-pillar (the pillar between the front and rear doors). This is equivalent to two identical vehicles having a full-frontal head-on collision.

Some four-wheel-drive vehicles (also called SUVs, or Sports Utility Vehicles) have frames that are too stiff. They can generate decelerations greater than 50 G in the 56-kph (35-mph) crash test. For example, the Kia Sportage peaked at about 60 G. Maybe NCAP should test an army tank to see if we can get G-forces in triple figures!

But the above figures are the decelerations experienced by the *vehicle* when it hits the barrier.

The crash test dummies (and you or I, if we are unlucky) will keep travelling forward and experience a *second* collision with the interior of the vehicle. Hopefully a combination of seat belt and airbag will cushion this collision and make it survivable.

Current United States Federal Standards insist the cars must be designed so that the crash test dummy will experience less than 60 G in a full-frontal collision into a solid barrier at 48 kph (30 mph). The G-force is measured at the chest of the dummy, when the deceleration lasts longer than 3 milliseconds (1 millisecond, or mS, is one thousandth of a second). For less than 3 mS, the human body can withstand considerably higher decelerations. Perhaps this is because the organs don't have time to start moving relative to one another and don't rupture or collide. (The *third* collision in a car crash is when the organs hit your bones.)

VARIATIONS ON THE THEME OF MURPHY'S LAW

The best laid plans o' mice an' men,
Gang aft agley [are likely to go wrong]
Robbie Burns, Scottish poet, 1786

- **Logic is a systematic method of coming to the wrong conclusions with complete confidence.**
- **The advantage of a computer is that it can make as many mistakes in two seconds as 20 people working for 20 years.**
- **When all else fails, read the instructions.**

THE SCIENCE OF MURPHY'S LAW

Robert A. J. Matthews wrote an article called "The Science of Murphy's Law" for the April issue of the *Scientific American* in 1997. Edward A. Murphy III, the son of Edward A. Murphy Jr. who proclaimed Murphy's Law, then wrote a letter thanking Matthews for his article, and for treating Murphy's Law seriously.

Calendars and posters make a joke of Murphy's law. But as Edward A. Murphy III wrote, "*Murphy's Law actually refers to the CERTAINTY of failure. It is a call for determining the likely causes of failure in advance and acting to prevent a problem before it occurs . . . Murphy and his fellow engineers . . . worked on supersonic jets and the Apollo landing craft . . . they knew that things left to chance would definitely fail, (so) they went to painstaking efforts to ensure success.*"

Murphy's Law of Queues?

If you look hard enough, you'll think that you can see Murphy's Law everywhere.

We've all come across Murphy's Law of Queues — you know, when you go to the supermarket, and there are 10 queues, why is *your* queue never the fastest?

Actually it's not really Murphy's Law, it just feels like it.

Over a period of time, on average, each queue will take its turn at being the fastest. But on any particular occasion, if there are 10 queues, your chances of being in the fastest queue are one in 10. So if you go to the supermarket 10 times, you'll probably be in the fastest moving queue only once. If you add in a bit of selective memory failure, and the fact that you almost certainly didn't take written notes of your shopping trips, you'll probably think that you are never in the fastest queue.

But in reality, averaged out over a period of time, you will get the fastest queue once out of every 10 visits — which is only fair.

Murphy's Law of Tumbling Toast?

But Murphy's Law of Tumbling Toast seems to pass the Reality Test. You reach across the breakfast table to top up your morning cup of tea. Your arm catches the toast, slides it to the edge of the table, and it tumbles off, always landing buttered side down.

Victorian Buttered Toast

The Tumbling Toast problem goes back a long way — it's not just you. Back in 1844, the Victorian poet and satirist James Payn wrote:

> I've never had a piece of toast
> particularly long and wide,
> but fell upon a sanded floor,
> and always on the buttered side.

Some people claim that the weight of the butter makes toast land buttered side down. But the weight of the butter is less than 10% of the total weight of a typical slice of

murphy's law on earth : the toast landing face down scenario

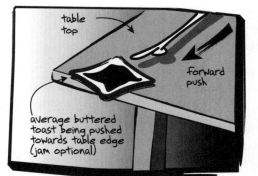

1. the toast & the shove

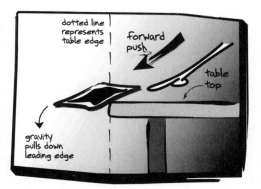

2. gravity begins to work

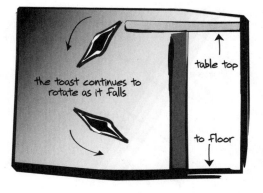

3. rotate, rotate...

4. the aftermath...

toast, and most of this butter gets sucked into the middle of the slice. The butter wouldn't have much effect upon the centre of gravity, or the rotational dynamics, of the tumbling toast.

Modern Buttered Toast

In 1995, Robert A. J. Matthews, a physicist who is also a science journalist, wrote a paper for the *European Journal of Physics* entitled "Tumbling Toast, Murphy's Law and the Fundamental Constants".

He first became interested in Murphy's Law in 1994, after he read a letter in the *New Scientist* claiming that every time a book slid off a desk, it would always land with the face that had been upwards now pointing down. In other words as it fell, it would rotate half a turn.

Do the Experiment

When Matthews tried the experiment, he saw the book beginning to rotate as its leading edge fell off the table. He realised that the letter writer was correct, and that, thanks to the height of the table, the book had time only to do half a turn.

So Matthews agreed that *"the popular view that toast falling off a table has an inherent tendency to land butter-side down is based in dynamical fact"*.

Matthews also realised that toast (or what he mathematically described as a *"rough, rigid, homogeneous rectangular lamina"*) landing buttered side down had very little to do with the extra weight of the thin layer of butter, or with any aerodynamic effects. The orientation of the toast was mostly controlled by gravity. The Earth's gravity pulls the leading edge of the toast down as it slides off the table, and starts it tumbling. If the Earth's gravitational field was stronger, the toast would rotate more than just half a turn.

Like any good physicist, Matthews combined theory and experiment. He compared both toasted and untoasted bread, and measured both the friction and the amount of overhang at the table's edge just as they were about to detach from the

cat fitted with breakfast table

BUTTERED TOAST AND TV SHOW'S BIZARRE EXPERIMENT

In 1991, the BBC TV series "QED" thought that they had disproved Murphy's Law about Tumbling Toast. They got volunteers to flip buttered toast up into the air, and see which way it landed. Out of 300 flips, 148 had the toast land buttered side up, while 152 landed buttered side down. This is pretty darn close to 50% — in other words, there was no real tendency to land one way or the other.

At first, they thought that Murphy's Law was acting on their experiments to make things go wrong — to make it look like Murphy's Law did not exist. But in reality, they had just done a really dumb experiment.

Under the usual circumstances at the breakfast table, the buttered toast will first be dragged off the edge of the table, begin to rotate, and then fall. But because the table is less than a metre high, the buttered toast will have time only for half a turn. So if it starts buttered side *up*, it will land buttered side *down*.

That's the real reason why toast lands buttered-side down; it only has time for half a turn.

But in the TV program, they didn't drag-and-drop the toast, they simply *flipped* 300 slices of buttered toast, which is nothing like the real-life situation.

table and enter free fall. He didn't find a lot of difference between toasted and untoasted bread. He concluded that, at the table's edge, "*the torque-induced rotation should dominate the dynamics of the falling toast*" — that is, the rotation that the toast picks up as it falls over the edge controls why toast lands buttered side down.

Fundamental Constants of the Universe

Matthews also realised that Murphy's Law was linked to the very fabric of the universe. Toast lands buttered side down because of events that happened at the birth of the universe.

Toast lands buttered side down because of the height of the table. Tables are the height they are because of the height of people.

But the formula that gives you the maximum height of people contains two of the Fundamental Physics Constants of the universe. One constant controls the strength of chemical bonds that hold the skull together. The other constant controls the strength of the Earth's gravitational field. These constants were locked in as soon as the universe popped into existence as a result of the Big Bang, some 15 billion years ago. So these fundamental constants control why buttered toast lands buttered side down — which brings us back to Murphy's Law.

G-FORCES IN FORMULA ONE RACING

A Williams driver, Heinz-Harald Frentzen, recorded 5.99 G while he was qualifying for the 1997 Imola Grand Prix (GP). To generate this, he had to push on the pedal with a "weight" of about 150 kg (330 lb) — which was made easier by his weighing six times more than normal.

Formula One cars generate up to 2.8 G as they brake, and 2.0 G as they accelerate. Just to be able to handle the weight of their head with a helmet on it, the drivers have to build up their upper bodies and necks with special exercises. Even so, 63% of GP drivers have symptoms of neck pain.

Racing car drivers have also survived incredibly high decelerations during high-speed crashes. Decelerations in excess of 100 G have been measured in these crashes, but, fortunately, only lasting several milliseconds.

FORMULA ONE CAR ON THE CEILING

A Formula One Grand Prix racing car can generate enough "weight" to drive on the ceiling!

Your average fully-loaded Formula One car weighs just over 550 kg (1212 lb). It has special aerodynamic curves that push it down onto the road so the tyres will stick better and it can corner more quickly. A Formula One car travelling at 240 kph (150 mph) generates around 1600 kg (3500 lb) of extra "downforce". So it's stuck to the road at around 3.9 times its own weight.

If you had a suitably curved road tunnel, you could enter the tunnel on the floor and gradually work your way around to the ceiling. You have to subtract the weight of the car, but at 240 kph (150 mph) the overall "weight" holding the Formula 1 car to the ceiling is 1.9 times its own weight.

These aerodynamic forces are enormous, and can create havoc when they get out of control. In the 1999 Le Mans 24-hour race, the CLR Mercedes-Benz suffered the phenomenon of "kiting" — i.e., air getting under the car. The CLR did three somersaults above the track, reaching a height of 10 metres (33 ft), before it slammed tail-first into trees some 30 metres (99 ft) away from the track. The driver miraculously survived.

Tumbling Toast Solution

If you're really clever, you can invent a machine to interfere with the local gravitational field around your breakfast table to fix the buttered toast problem, so that it lands buttered side up. Or, if you don't mind eating standing up, you can eat your breakfast off the top of the fridge, so that the buttered toast gets enough time to do a full rotation and land buttered side up.

Robert Matthews himself suggests a not-so-obvious solution. Once you notice your buttered toast heading for freedom, give it a "*smart swipe with the hand ... (This minimises) the amount of time that the toast is exposed to the gravitationally-induced torque ... by giving the toast a large relative velocity ... the toast will descend to the floor keeping the butter side uppermost ... according to Einstein, God is subtle, but He is not malicious. That may be so, but His influence on falling toast clearly leaves much to be desired.*"

Or then again, you could butter your toast on the other side . . .

G-FORCES SAG YOUR FACE

One "advantage" of high **G**-forces is that you can see how you will look in 30 years' time!

According to **RAAF** (Royal Australian Air Force) squadron leader **Dr David Neuman** from the **RAAF Institute of Aviation Medicine** in **South Australia**, just aim the nose of a **F/A–18 Hornet** at the sky and accelerate hard — and if you have a mirror, you can watch your face sag.

He also said that in a typical air combat between two **F/A–18s**, the pilots would be exposed to accelerations as high as **7 G**.

MURPHY'S LAW OF MAPS

When you're navigating with a map, you don't want to flip from page to page. You would prefer the place you're looking for to lie close to the dead centre of the map, so you could stay on one page.

But the so-called **Murphy's Law of Maps** says, "*If a place you're looking for can lie on the inconvenient parts of the maps, it will*". This is **NOT** Murphy's Law, it's straight geometry.

Let's say that the evil "Murphy Zones" are those parts of the map that are close to the edges. Suppose that the Murphy Zone has a width one tenth of the width of the entire map. Suppose also that your map covers two pages, with a border on each page.

Ten per cent doesn't sound like much, but the Murphy Zone snakes around the outside edge of the map — the big bit. If you do the maths, the Murphy Zone covers more than half the area of the map. So you have a better than 50% chance that any random point on the map will lie within the outer 10% of the map.

REFERENCES

Robert L. Forward, "Murphy Lives!", *Science*, February 1983, p 78.

Mike Gaines, "Who'd Fly a Superfighter?", *New Scientist*, No. 1881, 10 July 1993, pp 28–32.

Robert A. J. Matthews, "Tumbling Toast, Murphy's Law and the Fundamental Constants", *European Journal of Physics*, Vol. 18, 1995, pp 172–176.

Robert A. J. Matthews, "The Science of Murphy's Law", *Scientific American*, April 1997, pp 72–75.

Ian Stewart, "The Anthropomurphic Principle" (in the column "Mathematical Recreations"), *Scientific American*, December 1995, pp 86–87.

MURPHY'S LAW IN SPACE

Murphy's Law began in the aerospace industry, so it's perfectly natural that the law should apply in space.

At vast expense, in 1982 the Soviets inadvertently proved that Murphy's Law works on Venus, while the Americans showed that Murphy's Law works on Mars in 1983.

Venus

Venus, named after the Roman goddess of love, is the second planet from the Sun. It's just a fraction smaller than Earth. Venus takes 224 Earth days to orbit the Sun, but rotates on its own axis (relative to the stars) once every 243 Earth days. So the Venusian year is shorter than the Venusian day!

Another strange thing about Venus is that it rotates in a direction opposite to the Earth — so the Sun rises in the west and sets in the east.

Venus is wrapped in very thick clouds that have always stopped us from peeking through to see the surface. For a long time, we had no idea of the conditions on the surface of Venus. Some writers of science fiction and fantasy imagined that Venus had steamy tropical jungles inhabited by dinosaurs!

But modern science has proved that the conditions on Venus are terrible. The temperature is around 460°C, or 860°F (hot enough to melt the zinc off your tin roof, and the lead out of your plumbing and solder), and the pressure is around 90 Earth atmospheres (which is more than enough to give you a severe case of the bends). The atmosphere is carbon dioxide and sulphuric acid.

So it's pretty clear that Venus has a very hostile environment.

Surface of Venus

Even with all their technology, the Americans have never succeeded in overcoming the terrible conditions on Venus to send back colour pictures of the surface. The Russians (or Soviets, as they were back then) finally struck it lucky with Venera 13 and 14 (*Venera* is a Russian word meaning Venus). These spacecraft sent back colour pictures of the surface of Venus and, at the same time, proved that Murphy's Law works on that planet too!

Venera 13 was launched on 30 October 1981. Venera 14 was launched a few days later, on 4 November. After travelling through space, and inserting into orbit, Venera 13 landed on Venus on 1 March 1982, and following faithfully behind, Venera 14 landed on 5 March.

FIRST UNITED STATES/SOVIET SPACE COOPERATION

The very first time the Soviet Union and the United States officially cooperated in a mission to any of the planets was in the planning of Venera 13 and 14.

The American Pioneer Venus Orbiter spacecraft had been using radar to map the shape of the planet Venus since 1978. As a result, NASA had an idea of what lay under the thick clouds of Venus for the first time. The Americans passed on their geological knowledge to the Soviets, who selected two different landing sites with very different geology.

Venera 13 touched down to the east of a small highland region called Phoebe Regio (457°C, or 854°F, and 89 Earth atmospheres). Even though the spacecraft had been designed to last only half an hour, it survived four times longer: 127 minutes. In that time, it sent back eight colour panoramic pictures as part of its dozen experiments. Venera 14 landed in the lowlands, about 1000 km to the southeast of Venera 13 (465°C, or 869°F, and 94 Earth atmospheres). The Soviets did not say how long it survived, but it also sent back colour pictures and analyses of the rock.

Soviets on Venus

Because time was tight, the two spacecraft had to get straight to work with their experiments. One experiment was to take photos of the ground once the craft had landed. For this to happen, a lens cap (used to protect the lens during descent) had to pop off. Another experiment was to measure the compressibility of the Venusian soil, using a spring-loaded arm.

At that point Murphy's Law stepped in.

On each spacecraft, the lens cap popped off and landed on the ground. Then the spring-loaded arms popped out. On Venera 13, the arm sent back to Earth the "*physico-mechanical properties*" of the ground. But on Venera 14, the arm sent back the measurements, not of the Venusian surface, but of the compressibility of a lens cap.

Murphy's Law on Venus

With 500 million square kilometres of Venus to land in, the lens cap had to land on the exact spot the arm was going to sample!

murphy's law in space: VENERA 14 & "that" lens cap

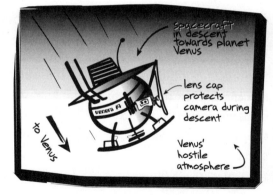

I. entry into the planet's hostile atmosphere

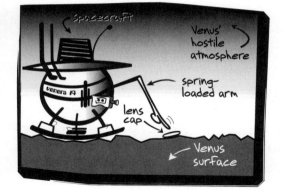

2. doh!! the lens cap!

As Captain Edward A. Murphy Jr. might have said, *"I guess that's the way the lens cap bounces".*

Mars

Mars, named after the Roman god of war, is also home to Murphy's Law.

Mars is the fourth planet from the Sun, and is about half the size of our Earth. Its day is 24 hours 41 minutes — only a little longer than Earth's. The Martian day is called a "sol", and there are 668.6 Martian sols (about 23 Earth months) in each Martian year.

The atmosphere of Mars is not breathable by humans (until we get into some serious genetic engineering). First, it's mostly carbon dioxide. Second, it's very thin — the pressure is only about 0.6% of that of the Earth. Even so, Mars has weather. Tornadoes, up to 6 km (4 miles) high, twist across the surface of the planet. Winds can last for months at speeds up to 350 kph (220 mph). Many of the canyons have early morning fogs.

The temperature is just too cold to be comfortable for us (more genetic engineering needed?). The Viking Landers measured temperatures between −123°C (−190°F) and −23°C (−10°F). The temperature varied by as much as 50C° (80F°) in a single day.

The Geography of Mars

Mars has the tallest mountain known in the solar system — Olympus Mons, about 27 km (17 miles) high, and 550 km (340 miles) across at the base. Mars has a huge system of parallel canyons that dwarfs the Grand Canyon. This system, the "Valles Marineris", is about 5000 km (3100 miles) long, plunges down to 7 km (4.3 miles) deep, and averages about 200 km (120 miles) wide.

Mars has two moons, named Phobos and Deimos, from the Greek words for terror and panic. Phobos is in an unstable orbit, and will crash onto the surface of Mars in about 30 million years. These two moons were written about *before* they were discovered with telescopes. Both Jonathan

Swift (in *Gulliver's Travels*), and Voltaire (in *Micromegas*) predicted the moons' sizes and orbits.

The Martian poles advance in winter and retreat in summer. They mostly consist of solid carbon dioxide ice, but the north Martian pole also has water ice. In fact, the water at the north Martian pole is probably the largest single reservoir of water on the surface of the planet. I personally think (but there is no evidence for this yet) that there is almost certainly water underground.

Life on Mars?

People have wanted to believe that there is life on Mars ever since the Italian astronomer and statesman Giovanni Virginio

MYSTERIOUS MOONS OF MARS

How did Voltaire and Swift come to predict that Mars had two moons, long before telescopes were good enough to see those moons?

Easy. They followed a mathematical progression.

The Earth had one Moon. Galileo had seen four moons around Jupiter (but by 1999 we had found about 16 moons).

Mars is between Earth and Jupiter. Therefore, they reasoned, Mars should have two moons, which is kind of in the middle between Earth's single moon, and Jupiter's four moons. So it was a lucky guess . . . twice.

Schiaparelli wrote about some "channels" that he claimed to have seen on the surface of the planet. Previous astronomers had described similar patterns, but his writings were the first to become really popular. From 1877 onwards, he documented over 100 "channels". Unfortunately, the Italian word for channel is *canali*, which the English-speaking media translated as "canals". Soon the speculation about the "Martians", their canals, and their supposed civilisation reached fever pitch.

Percival Lowell, from the United States, also claimed to see these "canals". He believed that the intelligent Martians had dug giant irrigation canals to carry water from the polar icecaps as their planet dried out. He thought that the "canals" he could see were bands of vegetation, many kilometres wide, on each side of the smaller irrigation canals.

But when the US Mariner 6 and 7 spacecraft sent back photos in 1967, we saw no such canals. They were merely an optical illusion, created when the eye of the astronomer "joined" individual large surface features to make what seemed like canals.

Liquid Water on Mars

However, the high-resolution photos from more recent spacecraft show us that there really are channels in certain parts of Mars. They don't look like an irrigation network, but they do look exactly like dried-up riverbeds. There is no liquid water on the surface of Mars today, but these channels prove that it must have been present millions of years ago.

The question is whether there is still underground water on Mars. If there is, life could survive in this water. After all, the further you go underground, the warmer it gets.

My guess is that the life would most probably take the form of some kind of bacteria. Future spacecraft will tell us if I'm right. What we know, however, is that Murphy's Law applies on Mars.

Vikings Set Sail

Back in 1975, two Viking spacecraft set sail for Mars and proved that even this far-off planet cannot escape the effects of Murphy's Law.

Each Viking was a two-part spacecraft, made up of an "Orbiter" and a "Lander".

Each Orbiter would orbit Mars, taking photographs. Between them, the two Viking Orbiters took some 52 000 pictures, covering about 97% of the surface of Mars. That 97% was photographed at 150–300-metre resolution, while selected areas were photographed at 8-metre resolution.

Each Lander would touch down on the surface to do science experiments, which included looking for life. The two Viking Landers sent back some 4500 pictures, and approximately three million weather measurements.

The Landers touched down in 1976, some 6500 km (4000 miles) away from each other. Because of the importance of surviving the landing, the two Viking Landers were targeted on open plains, rather than in the more interesting deep canyons.

Looking for Life

Each Lander had the same three specific goals in looking for life. First, they looked for organic debris on the surface. Second, they looked for anything that suggested the presence of living or fossilised organisms. Third, they did experiments to find an organism (e.g., bacteria) with a metabolism.

The results were tantalisingly vague at each site.

Life on Mars — Definitely Don't Know

First, the bad news. There was no organic debris found, according to the very sensitive chemical analyses. The cameras did not see any hints of life, past or present.

THE VIKING SPACECRAFT

At launch, each fully-fuelled Viking Orbiter–Lander pair weighed 3530 kg (7782 lb). After separation and landing, each Orbiter weighed 900 kg (1980 lb), while its Lander weighed about 600 kg (1320 lb).

Each Orbiter was powered by solar panels generating 620 W, which fed two 30-AmpHour nickel-cadmium rechargeable batteries.

Each Lander was powered by two radio-isotope thermal generator units (RTGs), containing plutonium-238. The two RTGs gave 70 W continuous power. Four 8-AmpHour nickel-cadmium batteries handled peak power loads. The Landers used nuclear power instead of solar panels, because an extended dust storm could block sunlight.

Lou's antenna repairs

THE VIKING JOURNEYS

Viking I was launched on 20 August 1975 and after a 10-month flight entered orbit around Mars on 19 June 1976. It immediately began looking for good landing sites. On 20 July 1976, the Viking I Lander separated from the Orbiter.

As it dropped through the thin atmosphere, it was slowed down by the wind resistance on a heat-resistant "aeroshell" on the bottom of Lander. At an altitude of 6 km (3.7 miles), when the Lander was travelling at about 900 kph (600 mph), the parachutes were opened. Within 45 seconds, the parachutes had slowed down the Lander to about 216 kph (135 mph).

The retro-rockets started firing at 1.5 km (0.9 miles) above the surface, and continued until the landing 40 seconds later at a speed of 8.6 kph (5.4 mph). The retro-rockets had a special 18-nozzle design, to spread the exhaust gases of hydrogen and nitrogen over a wide area. This design also meant that no more than 1 mm ($\frac{1}{25}$ in.) of the surface dust would be blown away, and that the surface would not be heated by more than 1°C (1.8°F).

The Viking I Lander touched down at the Plain of Chryse (*Chryse Planitia*). This is a rolling plain, strewn with boulders. The camera showed scattered dusty dunes between outcrops of bedrock, which are probably old volcanic lava flows. A locking pin on the soil sampler stuck, but was successfully shaken loose after five days.

Viking 2 was launched on 9 September 1975, and entered orbit around Mars on 7 August 1976. On 3 September 1976, the Viking 2 Lander also successfully touched down on a similar boulder-strewn area on the Plain of Utopia (*Utopia Planitia*).

Second, the good news. The experiment to look for signs of photosynthesis showed *"some of the signs of a positive result"*. And the Martian soil did give off oxygen when it was treated with organic nutrients.

All we can say with any certainty is that there are probably no living organisms on the surface in the two places we landed. But remember, the Viking spacecraft were directed to land in areas as flat as possible, so that they would not tip over. How much life would a spacecraft find if it landed in a supermarket parking lot here on Earth?

Age and Budget Cuts Attack Viking

Originally, the two Landers sent pictures back up to the two Orbiters, which then sent them back to Earth. But as some of the Orbiters/Landers died, the remaining ones were reconfigured.

In 1982, six years after the landing, only one of the four machines was still alive: the Viking 1 Lander. It sat on a plain that had been scoured by floods hundreds of millions of years ago. It had survived savage windstorms and freezing winters and was still a good little robot, sending back regular snap-shots and weather information directly from the Martian surface.

The politicians in Washington decided to save a few dollars, so they closed down the receiving station here on Earth. The American public responded generously and kept the Viking 1 in contact with us, by donating to what they called the Viking Fund. This money paid for a skeleton crew of workers from the Jet Propulsion Laboratory in California to keep the receiving station going.

Viking Limps On

Once a week, the Viking 1 Lander would swing its little antennae towards that tiny bit of sky where the Earth is, wait until it received a "go" message, and then tell us all the latest Mars weather news.

Can you see the weak link?

The Viking Lander computer might be loaded up with information for us, the Viking transmitter could be in terrific working order, but suppose that the Viking receiver wasn't working. In that case, it couldn't receive the "go" message, so it couldn't tell us the latest Mars weather.

The skeleton crew back on Earth decided to change the computer program so that the Viking 1 Lander would talk to us without having to receive a "go" signal. However, the computer had been built in 1971, and so it had a very small memory of only 6000 words. Whenever the crew wanted the computer to do something new, they had to wipe out part of the memory to insert their new instructions.

Murphy's Law Strikes

While they were loading up the new program, they accidentally wiped the location of Earth from the memory banks. The Viking 1 Lander didn't know where we were! It had no idea where to point the antenna.

The engineers tried everything.

They worked out where the new instructions would have made the antennae point. In January 1983, when they calculated that the rotation of Mars would have pointed the antennae directly at Earth, they tried to communicate with Viking 1. No luck.

Then they remembered the switch that shuts down the computer when the batteries are low. Perhaps, they thought, this switch may have triggered by itself. So on 16 February 1983, they tried to reset the switch. Again, no luck.

They even tried telling Viking 1 to switch on its back-up transmitter. Still no luck.

So maybe we can put intelligent life (if any) on Mars to the test. Until we can get to Mars, the Martians will be the only ones who can fix our lost Viking.

Murphy's Law in Space

In the year 2000, the city of Sydney will host the Olympic Games. All the communication links in and out of Australia will be strained to their maximum. Unfortunately, also around the year 2000, the Sun is expected to enter into another phase of violent activity, which could interfere with the Olympic broadcasts.

There's only one satellite that regularly monitors the Sun. Even though it was almost wiped out thanks to our old friend Murphy's Law, we managed to revive it.

Sun Fries Satellites

The Sun is a mighty beast. It can squirt out billion-tonne blobs of hot million-degree gas at one third the speed of light. The Sun squirts out several of these blobs every month, and they usually miss us. But if they do hit, you can kiss your communications satellites bye-bye.

Back in March 1989, at the previous peak of solar activity, the most violent magnetic storm for 30 years lashed the Earth. Not only did it destroy two communications satellites, it brought down the Hydro-Quebec power system in Canada. Since then, the Sun's activity has destroyed a few more communications satellites — the Anik E–1 and E–2 satellites in 1994, and the Telstar 401 in 1997.

SOHO Monitors Sun

The Solar and Heliospheric Observatory satellite (SOHO) was launched on 2 December 1995 to monitor the Sun. It was sent 1.5 million kilometres away from the Earth, towards the Sun. Here the

FATE OF THE VIKINGS

The Viking 2 Orbiter developed a leak in its attitude control gas (the gas which controlled the way the Orbiter was facing). This Orbiter was powered down on 25 July 1978, after 706 orbits.

The Viking 2 Lander was turned off on 11 April 1980, after the batteries failed.

By 7 August 1980, the Viking 1 Orbiter was running low on attitude control gas. It was put into a higher orbit, to stop it colliding with Mars (and possibly contaminating it) until the year 2019. It was fully shut down on 17 August 1980 after 1485 orbits.

The Viking 1 Lander operated until 13 November 1982, when "a faulty command sent by ground control resulted in loss of contact".

gravitational pull of the Earth exactly balances the gravitational pull of the Sun, so SOHO could look at the Sun continuously.

SOHO has sent back over two million images of the Sun and has made great discoveries. They include enormous rivers of hot gas flowing just under the surface of the Sun, and huge gyrating storms, which are similar to the tornadoes we have on Earth, only much bigger and faster, with gusts up to 500 000 kph (310 000 mph). SOHO found over 50 comets that narrowly missed the Sun, and some that actually fell into the Sun.

We Lose SOHO

But on 25 June 1998, it all went horribly wrong.

The ground crew received signals from SOHO that it had slipped into the Emergency Sun Reacquisition Mode (ESR). This is a "safe" or "self-protection" mode, where SOHO stops doing science and just concentrates on pointing its solar cells at the Sun. The ground crew had dealt with five ESRs before and wasn't worried. They started the recovery sequence to bring SOHO back under operator control. But before SOHO had fully recovered, suddenly a second ESR was triggered. Once again, the ground crew wasn't too concerned — a few months earlier in March 1998, SOHO had gone into a second ESR while recovering from a first ESR. But as the ground crew now tried to recover from the second ESR, a third one happened.

The operators didn't know it, but SOHO was now spinning out of control. The solar cells were not pointed at the Sun. Almost immediately SOHO lost power, the ability

to talk to Earth, and the ability to stay pointed at the Sun. Within 6 minutes it was silent, and lost.

Murphy's Law Attacks SOHO

How did this happen? Easy. First, there were two previously undetected errors written into the software that controlled the stabilising gyroscopes, which kept SOHO pointing at the Sun. These errors had caused the ESRs. Second, the ground operators made the really unfortunate decision to switch off the only remaining gyroscope that was trying to keep SOHO pointing at the Sun. Put them all together, and you have a lost SOHO.

So the engineers mounted a rescue mission.

They analysed the last hour of data sent from SOHO, and figured out that it was probably spinning with its solar cells virtually edge on to the Sun, not generating any power. But in three months, or a quarter of an orbit around the Sun, the panels would be pointing straight at the Sun. In a month or so, once the solar panels were catching a bit of sunlight, electrical power might possibly be restored.

We Find SOHO

On 23 July, the 305-metre-wide Arecibo radio telescope in Puerto Rico sent a 580-kilowatt signal to where the silent SOHO should have been. The NASA tracking dish in Goldstone, California, picked up a strong radar echo. This was a good start — even though SOHO wasn't working, at least it was still in one piece, and roughly where it was supposed to be.

MURPHY'S LAW POWERS UFOS (VERY SILLY)

What happens when you attach buttered toast to the back of a cat, and drop the toast/cat combination from a building?

The four feet of the cat want to hit the ground first, but so does the buttered toast. The theory goes that if you have the right amount of butter on the toast, it will exactly balance the tendency of the cat to land on its feet, and the cat will hover in the air, above the ground.

This explains how UFOs fly without external engines — they get their lift from hundreds of toast/cat pairs inside the UFO.

And the humming noise associated with UFOs? That's just the sound of hundreds of purring cats.

Meow!

Each day, the motion of SOHO around the Sun "turned" its solar panels one degree closer to the Sun. On 3 August, the NASA station at Canberra broadcast signals to SOHO, and, for the first time, SOHO replied in tiny bursts lasting 2–10 seconds, as the solar panels wheeled in and out of the sunlight. The ground crew soon found out that some of SOHO's internal components had chilled down to −60°C, while part of the outside had baked up to +80°C. Amazingly, the instruments survived these terrible temperature extremes.

We Reactivate SOHO

Gradually, the recovery team reconfigured the spacecraft to use electricity from the solar cells to reheat the frozen rocket fuel and recharge the batteries. By 16 September, they had been able to thaw out enough fuel to get SOHO facing the right way towards the Sun. By late October, all of the instruments were back in operation. In fact, one instrument detector was performing better than before, thanks to having been baked by the Sun.

Unfortunately, two out of three gyroscopes had been permanently damaged by the intense cold.

On 21 December, the remaining gyroscope failed. This was a major disaster. SOHO had only 180 kg of fuel left. Using the rocket thrusters to keep SOHO facing the Sun would use up 7 kg of fuel each week, running it dry within about 25 weeks. But the engineers came up with a clever solution. They reprogrammed the spacecraft computers so that it would keep itself pointing at the Sun via its internal reaction control wheels.

At the moment, SOHO is working just fine, and should continue operating up to 2003 — well beyond the solar peak around 2000. If SOHO keeps operating, it will warn us when the Sun hurls a giant blob of hot gas at us, so that we can shut down the communications satellites for an hour or so, and then get back to broadcasting the Olympics.

Joseph C. Anselmo, "Controller Error Lost Soho", *Aviation Week and Space Technology*, 27 July 1998, p 28.

Leonard David, "Saving SOHO", *Aerospace America*, May 1999, pp 60–67.

Nigel Henbest, "127 Minutes Under Venus's Orange Skies", *New Scientist*, No. 1296, 11 March 1982, pp 623–624.

"Look Homeward, *Viking 1*", *Science 83*, April 1983, p 7.

Dennis Overbye, "Back to the Planets", *Discover*, May 1983, pp 28–33.

"Sunset for Viking", *Discover*, July 1983, pp 8, 10.

Michael A. Taverna, "Panel Urges Major Changes in Soho Ground Operations", *Aviation Week and Space Technology*, 14 September 1998, pp 58–59.

"That's the Way the Lens Cap Bounces", *Science 82*, September 1982, p 7.

FAKE FLIES AND CHEATING CHEETAHS

Faster than a falling falcon? Not likely, when that falcon is the Peregrine Falcon, which can reach 349 kph (217 mph) as it dives for the kill. That speed gives it the title of Fastest Animal on Earth. It is also claimed that the fastest mammal is the cheetah at 112 kph (70 mph), and that the fastest insect is the Deer Botfly at 1317 kph (818 mph) — which, by the way, is faster than the speed of sound!

However, it seems that someone has been cheating on the cheetah figures, and that the stats for the botfly have been badly botched. But once any record — right or wrong — is published, it takes a long time to set the record straight.

Supersonic Insect

The botfly (*Cephenemyia pratti*) story began back in 1927. It was then that the entomologist Charles H. T. Townsend published an article in the *Journal of the New York Entomological Society*.

It's a curious article.

He was fascinated with the idea that one day aeroplanes might be able to fly so rapidly that they could travel at the speed of the sunrise over the Earth. Of course, the longest path would be on the equator, while the shortest would be at one of the poles. At somewhere in between, such as the latitude of New York, he says that travelling at 1311 kph (815 mph) would *"exceed the speed of the dawn"*. (It was only on 14 October 1947 that Captain Charles "Chuck" Yeager broke the sound barrier in the Bell X–1, with a speed of 1078 kph or 670 mph).

How would we humans ever build such a fantastic aeroplane? By copying the magnificent flight mechanisms of his favourite insect, the deer botfly, says Townsend. He then spends five pages lovingly describing, in exquisite and intimate detail, all that he knows about the *"peculiar structure of the fly wing"*. He was convinced that if we designed a new plane, we should copy the fly, not the bird.

Townsend writes about the "amazing" speed of the deer botfly: *"Regarding the speeds of* Cephenemyia, *the idea of a fly overtaking a bullet is a painful mental pill to swallow, as a friend has quaintly written me, yet these flies can probably do that to an old-fashioned musket ball. They could probably have kept up with the shells that the German Big-Bertha shot into Paris during the world war. The males are faster than the females, since they must overtake the latter for coition."*

Seen with Own Eyes

Townsend then writes about how, when he was on top of a 2133-metre (7000-ft) mountain, he had seen pregnant botflies *"pass while on the search for hosts at a velocity of well over 300 yards per second — allowing a slight perception of colour and form, but only a blurred glimpse"*. (I'd be very impressed if he could tell you that a *"blurred glimpse"* was a specific fly, let alone a pregnant one!) This speed works out to an astonishing 987 kph (614 mph). He believed that botflies at higher altitudes would encounter less air and generate less air friction, so they could go faster.

Townsend continues, *"On 12 000-foot summits in New Mexico I have seen pass me at an incredible velocity what were quite certainly the males of (the deer botfly). I could barely distinguish that something had passed — only a brownish blur in the air of about the right size for these flies and without sense of form. As closely as I can estimate, their speed must have approximated 400 yards per second."* This works out to 1317 kph (818 mph).

He concludes his article by pointing out that insects are *"more efficient"* than birds, and that aeroplanes based on his favourite insect *"should far excel anything heretofore accomplished in the air"*. Back in 1927, nobody knew what the best design for an aeroplane might be. Townsend followed the old rule of *"Don't reinvent the wheel,"* and nominated his beloved deer botfly as a superb design.

But apparently, he never actually tried to *measure* the speed of the botfly, and at no time did he use any scientific apparatus. He merely *guessed* the speed by looking at the blurs.

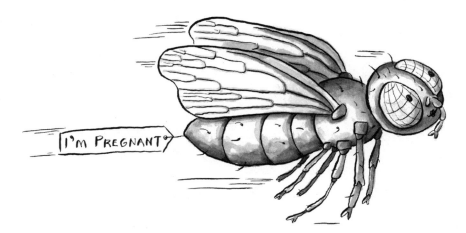

I'M PREGNANT

BOTFLY FACTS

Botflies are just hairy flies. Their special trick is that their larvae live inside the bodies of mammals as parasites. Different types of botfly attack different species.

The horse botfly uses an irritating glue to stick its eggs to the body hairs of a horse, donkey or mule. When the animal tries to lick off this itchy glue, it also swallows the eggs. They grow inside the stomach until they are ready to pupate. They then leave the stomach and are carried through the digestive tract of the horse until they drop out with the faeces. There are also sheep botflies and cattle botflies.

The human botfly of Central and South America is particularly ingenious. It captures a mosquito and attaches its eggs to it. Then it releases the mosquito. The mosquito bites a human, the eggs hatch, and the larvae migrate into the human. They drill in until they reach muscle. They then grow bigger and eventually drill out to continue the reproduction process.

The "Fact" Spreads

Townsend's article was followed up by an editorial in the *New York Times*, warning humankind not to be too proud of the new speed record of over 482 kph (300 mph) for a seaplane, because the deer botfly could easily beat that with a speed of 1126 kph (700 mph). This "fact" about the speedy botfly rapidly made its way into the hearts and minds of the public.

On 1 January 1938, the *Illustrated London News* quoted Townsend's claims again. It said that the female botfly could reach 988 kph (614 mph), but that the male could zip along at 1317 kph (818 mph).

Real Science

Irving Langmuir was an American chemist and Nobel Prize winner. He had a good grasp of basic science and knew the importance of actual measurements. He found Townsend's astounding claim impossible to believe. He wrote an article in *Science*, where he demolished the "Supersonic Botfly" assertion. Langmuir came up with three theoretical objections and one experimental objection.

First, Langmuir used well-known equations for wind resistance. He calculated that at 1317 kph (818 mph) the air pressure against the head of the fly would have been about half an atmosphere. This pressure of about 5 tonnes per square metre (4.1 tons per square yard) would have been enough to crush the poor little fly.

Second, he used the same equations to work out the power needed to overcome the wind resistance at that speed. The deer botfly would have to generate half a horsepower. Think about how big a horse is, and how big its muscles have to be to generate one horsepower. It seems

According to the *Guinness Book of Records*, the cheetah probably holds the ground speed record of around 100 kph (62 mph) over a short distance of up to 550 metres (1800 ft).

But the long-distance (815-metre or 3000-ft) ground speed record probably goes to the Pronghorn Antelope of the western United States. They have been clocked at 88 kph over 800 metres (55 mph, ½ mile), 68 kph over 1.6 km (42 mph, 1 mile) and 56 kph over 6.4 km (35 mph, 4 miles).

Milton Hildebrand reports that "*Antelopes apparently enjoy running beside a moving vehicle; they have been reliably timed at 60 mph (96 kph)*".

It would be ironic if the antelope turned out to be faster than the cheetah.

impossible that a creature much smaller than the eyeball of a horse could generate half a horsepower!

Another effect of the enormous wind resistance would be that it would generate enough heat to fry the little botfly.

Third, where would this energy come from? From food, of course — a *huge* amount of food. Langmuir calculated that the botfly would have to eat one-and-a-half times its own weight in food every second!

Real Measurements

At this stage, Langmuir did what Townsend should have done. He took some real measurements. They were a little rough, but they were, as scientists like to say, in the right "ballpark".

He decided to check at what speed a flying insect does turn into a blur.

This was ridiculously easy to test.

Langmuir got a lump of solder the size of a deer botfly ("*about 1 cm long and 0.5 cm in diameter*"), tied it to a light silk thread and swung it in a vertical plane. At

around 21 kph (13 mph), it was a blur: "*the shape could not be seen, but it could be recognised as a small object of about the correct size. At 26 mph (41 kph) the 'fly' was barely visible as a moving object. At 43 mph (69 kph) it appeared as a very faint line and the direction of rotation could not be recognised. At 64 mph (103 kph) the moving object was wholly invisible.*"

Langmuir then worked out that at the quite reasonable flying speed of 40 kph (25 mph), the botfly would probably need to eat food equal to only 5% of its body weight each hour. He concludes, "*A speed of 25 mph is a reasonable one for the deer fly, while 800 mph is utterly impossible*".

He mentions another very good argument against a botfly travelling faster than the speed of sound. A supersonic botfly smashing into you would make a big hole — but there are no such injuries on record.

The fastest insect on record that has been reliably measured is the Australian dragonfly (*Austrophlebia costalis*). In a few short bursts, it has been clocked at around 58 kph (36 mph).

So if the deer botfly is definitely not a supersonic insect, how swift is the cheetah? One thing that I have realised from reading the literature is that the speed of the cheetah is very famous but overall very badly measured.

Cheetah Extinction

The cheetah is very fast. But it might not be fast enough to outrun its own extinction.

Cheetahs of at least four different subspecies once roamed through North America, Asia, Europe and Africa. In fact, the name "cheetah" comes from the Hindi word meaning "the spotted one". Today they exist in sub-Saharan Africa as only one subspecies (*Acinonyx jubatus*), with a tiny remnant population barely surviving in northern Iran.

We're not sure of the exact numbers, but we do know that cheetahs are endangered.

ODD CHEETAH

Cheetahs are the only members of the big cat family that cannot roar. Instead, they purr.

In the mid-1950s, there were 20–40 000 cheetahs in the whole world. By the mid-1970s, their population had halved. According to J. G. Caughley, "*how far they dropped further between 1975 and now is anybody's guess*". Their main problem is that people kill them and take over their land so the surviving cheetahs have nowhere to hunt and live.

Superb Hunter

The cheetahs have been domesticated more than any of the other big cats.

CHEETAH BABIES

After a gestation of three months, the mother cheetah gives birth to a litter of two to six cubs. For the first three months of life, the cub has a distinctive dark spotted coat with long blue-grey hair on its back, neck and head, and darker fur underneath.

Around four months of age, this changes to the adult's coarse fur, which is white on the underbelly and tawny yellow on the back. There are many black spots over the fur, and a black stripe (the "tear line") runs down from each eye. A typical adult cheetah stands about 80 cm (31 in.) high at the shoulder, is about 140 cm (55 in.) long, and weighs about 55 kg (121 lb).

In the Serengeti, over half of all cheetah cubs die before they reach three months of age. They are defenceless when their mother leaves them to go hunting for food. The mother knows this, and will usually shift the cubs to a new lair every second day. But if they can survive their dangerous infancy, they can live for as long as 14 years in the wild. By 15 months, they are as big as their mother, and are as good as she is at chasing and killing animals.

The Sumerians, way back in 3000 BC, were the first to use cheetahs as hunting companions. Since then, the pharaohs of Egypt, the kings of France, the princes of Persia, the Mongol emperors of India and the emperors of Austria have continued this tradition. When Marco Polo visited Kublai Khan at his summer residence in the Himalayan Mountains 700 years ago, he found that mighty Khan kept 1000 cheetahs to hunt deer and other slower animals.

The cheetah is superbly engineered. It has a small light skull and long legs. It has a huge heart and oversized lungs, blood vessels and adrenal glands. The cheetah is the only cat that cannot retract its front claws fully, so they always protrude slightly. This gives them a better grip on the ground. It has a deep chest and slim waist, just like a greyhound. When a cheetah runs at full speed, the spine expands and contracts in each stride. It lengthens to let it reach further forward with its front legs, and shortens to give it better acceleration with the back legs.

The result is that the cheetah can probably accelerate and run faster than any other animal in the world .

Not So Good a Killer

Cheetahs tend to hunt in the early morning or the late afternoon. When they hunt alone, they mostly attack animals smaller than themselves. On the rare occasions that they hunt in a group, they will hunt bigger prey, such as a zebra.

CHEETAHS INBRED

Cheetahs are very inbred (bred from closely related parents). They are so inbred that they are almost genetically identical.

The current theory is that they became inbred when a "natural" disaster dropped their global population down to less than seven individual cheetahs. This probably occurred about 10 000 years ago. They then went through a "genetic bottleneck", and their genetic diversity plummeted. They survived only through brother-to-sister or parent-to-child mating.

If a species does not have much genetic diversity, it will not be able to adapt well to changes in its environment, such as climate change, new bacteria or viruses. But if they do have a lot of genetic diversity from one individual to the next, at least a few of them will be able to survive the changing times.

Tests involving *enzymes, skin grafts and skull shape* make us think the cheetahs are inbred. The enzyme tests probably give the strongest evidence for inbreeding. The tests involving skin grafts and skull shape provide weaker evidence.

Enzymes are medium-sized proteins that speed up chemical reactions (they are advertised in some washing powders). In your body, enzymes speed up the burning of food for energy by about one million times. (The average human life span is less than 1 000 000 hours. It takes at least an hour to digest a meal. So if you didn't have enzymes in your gut, it would take you an entire lifetime to digest your first meal!)

You have many thousands of different enzymes in your body. A typical enzyme is made up of, say, 100 or so amino acids. In any one enzyme, about 98 of the amino acids will be common to all humans, but two of them will vary from person to person. Even though there's a slight variation, that enzyme will still work perfectly well in all humans. If you go to the trouble to analyse the amino acids in, for example, 200 different enzymes in many different people, you'll find that 140 of these 200 different enzymes are completely identical in all their amino acids, and that 60 have minor differences that don't affect how they work.

So, according to the enzymes, humans rate at about 70% identical. But laboratory rats and cheetahs rate at 97% identical. Laboratory rats have been inbred via brother-to-sister mating for at least 20 generations. So cheetahs are at least as inbred as laboratory rats.

In a *skin graft*, you transplant skin from one place to another. If you are burnt on your face, the surgeon may graft on some skin from your legs or a buttock. The operation will usually be a success, because your immune system won't reject your own skin. However, you will almost always reject a skin graft from another person unless they are your identical twin. But about 50% of the time, cheetahs will accept skin grafts from each other rather than rejecting them. This means cheetahs must be genetically very similar to each other.

Cheetah *skulls* are not symmetrical. The scientists examined east African cheetah skulls, currently held in American museums. (Many of these skulls were collected by the U.S. President Theodore Roosevelt.) The left side of each skull is different from the right side, and is not a mirror image. We don't know why, but, in general, the more inbred an animal is, the more asymmetrical the skull is.

So the evidence points to the likelihood that they are very inbred. But why do some scientists think that cheetahs were reduced to a population of fewer than seven individuals, about 10–12 000 years ago?

They think *fewer than seven individuals existed* because it has been shown that if a population is reduced to seven individuals and then expands quickly, the offspring still retain about 95% of their genetic variability. But cheetahs have almost zero genetic variability (hardly any difference between them), so the numbers must have fallen below the magic "seven".

They think this took place *about 10–12 000 years ago* because back then, we know there was a massive destruction of many different mammalian species, such as mammoths, sable tigers and cave bears. About 75% of all mammalian species died out in North and South America. So this was probably the "disaster" that knocked off most of the cheetahs. Perhaps this disaster was a severe climate change associated with the tailing-off of the last Ice Age.

Whatever the cause, some scientists think that the cheetahs were almost totally wiped out — perhaps more than once. Like true copycats, they then built up their numbers with generation after generation of brother-sister inbreeding.

Once again, cheetahs are very endangered, but this time it's because of what we humans are doing to them. Now the Hindi name for the "cheetah" is the "spotted one" — and sure enough, if we keep killing them, you'd have to be very lucky if you spotted one.

Once they reach their prey, they usually knock it down with a powerful swipe from a front paw, or trip it over. The cheetah will usually kill the animal as soon as it is down, by biting its neck and strangling the windpipe.

Cheetahs have a reputation for speed and efficiency in killing. But they are not as successful at killing as most of the other African predatory carnivores, such as the lion and the spotted hyena. In fact, the only predatory carnivore that they consistently outrank is the jackal.

But Only a Sprinter . . .

Cheetahs are very good at chasing, so long as it's over a short distance.

In a sprint, their breathing rate increases from 60 to 150 breaths per minute. They have to be very careful stalking their prey, so that they can finish with a short sprint of less than 300 metres. Even this short a run at high speed exhausts them. After a chase, the cheetah takes half an hour to regain its strength. In this half-hour, the tables can turn, and the cheetah and even its young can be killed. If the prey is able to quickly scramble to its feet and run, the cheetah will usually let it go. In fact, cheetahs are so weakened by the sprint that in 50% of kills, their hard-won meal is stolen from them by leopards, lions and hyenas during this 30-minute window.

Some observers claim that 600 metres is the maximum distance ever seen for a prolonged chase by a cheetah. One author wrote about how two (much slower) mongrel dogs were able to corner a cheetah after relentlessly following it for 4 km (2 ½ miles).

Even though it's claimed that the cheetah can sprint at 112 kph (70 mph), the average final sprint is almost always done at less than 65 kph (40 mph).

Cheating Cheetah

The claim that the cheetah can travel at 112 kph (70 mph) originally came from an article by Kurt Severin published in the magazine *Outdoor Life* in April 1957. He wrote about a specific cheetah called Ocala.

Ocala's trainer and owner was John Hamlet. Hamlet had a master's degree in zoology. He owned and ran the Birds of Prey Wildlife Center near the town of Ocala in Florida. Previously, he had worked with the United States Fish and Wildlife Service, and had also headed a research centre for the National Foundation for Infantile Paralysis. He had played a part in developing the famous Salk polio vaccine, by capturing 2600 monkeys in the Philippines and shipping them back to the United States.

There aren't too many books on How to Tame a Cheetah, so Hamlet mostly worked out his own method. Part of Hamlet's taming program was getting Ocala "used" to his presence. He did this by reading to the cheetah — two Agatha Christie whodunits, and one Mike Shayne detective story.

First Cheetah Speed Experiment

To work out Ocala's speed, John Hamlet marked out a course 80 yards (73 metres) long. He attached a bag that was scented with meat to the wheel of an upside-down bicycle, and then hand-cranked the pedals like crazy. John Hamlet called out "GO!", started his

CHEETAHS INFERTILE

Various dictatorships and monarchies have given us a long tradition of keeping cheetahs as pets, hunting companions and status symbols. Even though this history spans 4000 years and covers three continents, until recently there was only one single case of successful breeding of cheetahs in captivity. This was done by Akbar the Great, a 16th-century ruler in India, who kept 1000 cheetahs. It wasn't easy. He had to let the cheetahs run free in his extensive palace gardens.

The first modern case of successful breeding of cheetahs in captivity happened only as recently as 1956 in the Philadelphia Zoo. Since then, we have learnt a lot about breeding cheetahs in captivity.

South African scientists discovered that the sperm count of the cheetah is pretty low. (Sperm count is the number of sperm per millilitre. In humans, the sperm count is between 40 and 200 million per millilitre.) The cheetah sperm count is 10 times lower than the sperm count of the lion, tiger or even the domestic cat. Also, about 75% of the sperm are damaged, and can't fertilise an egg. (The equivalent figure for the domestic cat is about 30%.) This makes cheetahs lousy breeders, but it doesn't have any effect on their love life.

Some people say that there is a link between animals being inbred and difficulty in conceiving. This may be true in most cases. But one exception is the Père David deer. Even though it is very inbred, having shrunk down to a very small group, it has since grown to many separate healthy populations.

stopwatch, and Ocala took off, chasing the meat-smelling bag. An assistant fired a pistol when the cheetah crossed the finish line, at which point Hamlet stopped his stopwatch. On the occasion that Severin was there, Ocala would run only twice. He wrote that *"from a deep crouch Ocala spurted to the end of the 80-yard course in 2¼ seconds, for an average speed of about 71 mph"*. That figure of 71 mph works out to 114 kph.

Many zoos and encyclopaedias repeat this figure. A typical statement from a large zoo reads, *"Cheetahs are the fastest animals on land — able to reach speeds of up to 70 miles per hour for short bursts!"*

Cheetah Speed Dispute Begins

But in November 1959, Milton Hildebrand wrote an article for the *Journal of Mammology*. He found three main errors in Severin's article that cast doubt on the claim that cheetahs could travel at 114 kph (71 mph).

First, there was a simple mathematical error! Eighty yards in 2.25 seconds works out to 117 kph (72.7 mph), not 114 kph (71 mph).

Second, they measured the track wrongly. The track was only 65 yards, not

CHEETAH LIVERS

In the wild, the cheetah hardly ever has liver disease, but in captivity, it's very common — up to 60%.

There are two theories explaining why this happens: *feminising hormones in the diet* and *lack of exercise*.

The first theory blames their diet. In the wild, cheetahs eat whole carcasses. In captivity, especially in North American zoos, they usually eat horsemeat mixed with soybean products to give extra protein. The soybean brings with it two natural chemicals, diadzein and genistein, which behave like weak oestrogens (female sex hormones). They interfere with the normal operation of the liver and can cause liver disease. Certainly, when the diet is changed to exclude soy products, the rate of liver disease drops.

The second theory says that the cheetah liver has adapted to the difficult job of providing lots of energy in a big rush. When the cheetah doesn't get much exercise, the liver deteriorates.

80 yards long. This slows down the 117 kph to 95 kph (59 mph).

Third, there's a lot of room for human reaction time error in measuring a time as short as 2¼ seconds. If the time had been longer, the reaction time errors would have been less important.

How they could work out any calculations from their data, let alone the cheetah's top speed, is hard to guess.

Of course, there was a fourth error. The animal was timed from a *standing* start, *not* once it had got up to speed.

Even today, we don't have a really good measurement on how fast a cheetah can run.

Second Cheetah Speed Experiment

The best measurement that we do have was made way back in 1965 by N. C. Craig Sharp, who was at that time a veterinary surgeon in Kenya, and who is currently the Professor of Sport Sciences at Brunel University in England. He had access to a tame orphaned cheetah that had been raised on a farm.

Sharp wanted to see how fast this cheetah could run, so he measured out a 201-metre (220-yard) course. He did this accurately, by using a surveyor's tape measure. He marked out the start line with a taut length of white wool, which he could easily see from 75 metres away. He drew a clear line between the posts at the finish.

He used an Omega analogue stopwatch, which was accurate to 0.1 second, and which had been calibrated against four other similar stopwatches. The day was so still that *"there was insufficient wind to move wool hanging vertically from a post"*.

Sharp was sensible enough to realise that the animal needed a bit of time to reach top speed, so the cheetah started off some 18 metres (20 yards) behind the starting line. Sharp was in the back of a Land Rover, trailing a lump of meat in the path of the cheetah. The Land Rover was some 75 metres away from the start posts. The vehicle took off with the meaty lure, and the cheetah chased. Sharp simply clicked his stopwatch on when he saw the cheetah break the white wool, and he clicked it off when he saw the chest of the cheetah cross the finish posts. He threw the meat to the cheetah at the finish post.

Cheetah Speed Results

Sharp got the cheetah to run the course three times. The first and third runs were done in one direction, with the second run in the other direction. Three runs is not large enough to be a statistically significant sample size. Even so, he did measure times of 7.0, 6.9 and 7.2 seconds. This works out to around 103 kph (64 mph).

At the time, Sharp didn't bother publishing his results, because he thought that other people had done more scientifically rigorous measurements of the speed of a running cheetah. He didn't think that his amateur results were the most accurate that had ever been done. But when he did eventually realise the gap in scientific research into this area, he released his results in 1997, in the *Journal of Zoology*.

We have seen a long history of incredibly rickety results. How can we possibly believe that the cockroach has a top speed of only 2.9 kph (1.8 mph)? They must be much faster, to get out of your way so rapidly when you enter the kitchen at midnight!

Sue Armstrong, "Happy Hunting Keeps Captive Cheetahs in Shape", *New Scientist*, No. 2050, 5 October 1996, p 22.

Milton Hildebrand, "Motions of the Running Cheetah and Horse", *Journal of Mammalogy*, Vol. 40, No. 4, 20 November 1959, pp 481–494.

Irving Langmuir, "The Speed of the Deer Fly", *Science*, 11 March 1938, pp 233–234.

Mike May, "Speed Demons", *The Sciences*, January–February 1999, pp 16–18.

Stephen J. O'Brien, David E. Wildt and Mitchell Bush, "The Cheetah in Genetic Peril", *Scientific American*, May 1986, pp 68–76.

Kurt Severin, "Speed Demon", *Outdoor Life*, Vol. 119, April 1957, pp 54–56, 81–83, 117–118.

N.C.C. Sharp, "Timed Running Speed of a Cheetah (*Acinonyx jubatus*)", *Journal of Zoology (London)*, Vol. 241, 1997, pp 493–494.

Charles H.T. Townsend, "On the *Cephenemyia* Mechanism and the Daylight-Day Circuit of the Earth by Flight", *Journal of New York Entomological Society*, Vol. XXXV, September 1927, pp 245–252.

"Zoo Diet to Blame for Cheetah's Sterility", *New Scientist*, No. 1580, 1 October 1987, p 31.

REFERENCES

KEVIN BACON SYNDROME

A few years back, for a big night out people played a party game called "Six Degrees of Kevin Bacon". In this game you had to link Kevin Bacon to any other actor, via shared movies or co-stars.

It was a fun party game, and, believe it or not, it was also hugely helpful for the Advancement of Scientific Knowledge. When a few scientists looked closely at the mathematics behind the "Six Degrees of Kevin Bacon", they uncovered patterns that could reduce the numbers of sexually transmitted diseases and improve your phone network.

History of the Kevin Bacon Game

The game was invented in 1993 by University of Pennsylvania students Mike Ginelli, Craig Fass and Brian Turtle. Apparently they had too much time on their hands. They soon appeared on television with their game, which became so popular that it swept across college campuses and the Net. It turned into a board game, and then the book *Six Degrees of Kevin Bacon* (with a bright green cover, and an introduction written by the star himself).

Suppose you wanted to link Kevin Bacon and Harrison Ford. Kevin Bacon has never worked directly with Harrison Ford in a movie. But Kevin Bacon and Lawrence Fishburne were co-stars in the movie *Quicksilver* — that's one link. Fishburne and Sean Connery worked together in the movie *Just Close* — that's the second link. And Connery and Harrison Ford worked together in *Indiana Jones and the Last Crusade*. So Ford and Kevin Bacon are only three links, or degrees, apart.

By the way, the six "degrees" means six "steps" between people. Six "degrees" does not mean six people including the

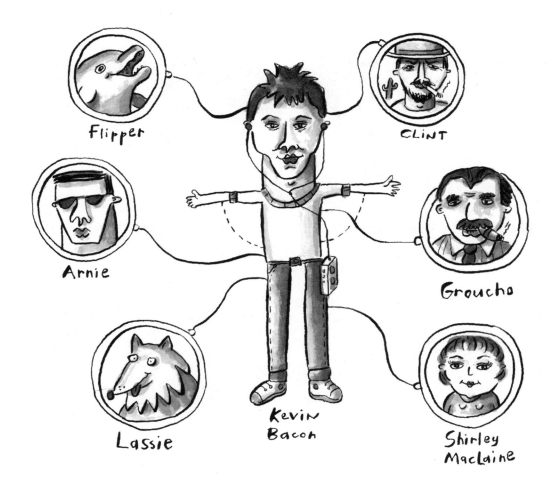

Flipper

CLiNT

Arnie

Groucho

Lassie

Kevin
Bacon

Shirley
Maclaine

nominated person and Kevin Bacon. So Harrison Ford and Kevin Bacon are three degrees apart, and there are four people involved in that particular chain.

You can assign any actor a "Bacon Number", which is the number of degrees of separation between them and Kevin Bacon. About 90% of all American movie actors have a Bacon Number of four or smaller.

Why Kevin Bacon?

Why did they pick Kevin Bacon — besides the fact that "Kevin Bacon" kind of rhymes with "separation" and has the same rhythm?

They "guessed" that Kevin Bacon is the centre of the movie universe. He may never

have been nominated for an Oscar, and he is not the biggest box office drawcard — but he has been consistently making movies since 1978. Some of his 30-plus movies include *Flatliners*, *The Air Up There*, *The Big Picture*, *Footloose*, *The River Wild*, *JFK*, *Animal House* and *Apollo 13*. There are many sides of Bacon.

In fact, he has so many sides that you can link him to any other American actor in the last century in six or fewer links.

Rules of the Kevin Bacon Game

There are no "official" rules to "Six Degrees of Kevin Bacon", because they are passed on by word of mouth — and usually at a party.

the Kevin Bacon syndrome

an "organised network" is similar to playing checkers

the rules of the game are such that you can only move one space at a time... (except when you are jumping or "crowned" and can move in any direction BUT still only one space at a time)

black

white

1. organised network - checker board model

a "random network" is similar to playing snakes & ladders

normally you move only one space at a time. but at a snake or ladder you can be whisked forward or backward many spaces.

2. random network - snakes & ladders model

There are, however, a few different ways to bring home the bacon in this game. There are even search engines on the Net that take all the fun out of it by giving you a whole bunch of answers.

You might try to link Kevin Bacon to another actor in the fewest possible steps, or in exactly six steps, or using only mainstream movies, or with very obscure and complicated pathways. Some rules allow short cameo roles, and others don't. But practically nobody will allow just a voice-over to count as a link — the person actually has to appear on screen.

Most people play this game face to face, but some (such as the Baconsortium) play it over the Net. The Baconsortium players are fairly rigid and won't allow obscure foreign language movies, TV shows, TV mini-series or telemovies. They certainly won't allow links through directors, writers, producers or relatives.

Everybody's Linked to Everybody Else

You might have heard somebody say at a party that they're linked to anybody else in the world by only five or six intermediate

SIX DEGREES OF SEPARATION

John Guare wrote a play (and later a movie script) called _Six Degrees of Separation_.

The story begins with a married couple (rich New York City art dealers) meeting a charming young con man. He convinces them not only that he is the son of the famous actor Sidney Poitier, but that he also knows their children who are studying at university. They give him a room for the night, and he cooks them a delicious meal. But the next morning, it becomes clear that the young man is not quite what he claims to be. The young man helps the couple recognise, and abandon, the emptiness of their life.

Part of the philosophy behind the play (and movie) is that everybody is related to everybody else in the world through just six other people, or six events or circumstances.

people. And you have probably met a stranger who turned out to know one of your friends, so you were truly linked to this stranger by a few degrees of separation. This linking sounds reasonable — at first.

Let's start off by supposing that everybody on the planet knows 100 people. You have 100 friends, so you are one degree away from 100 of the people on the planet. Now each of your friends knows 100 other people. One hundred multiplied by 100 is 10 000 — so already you are two degrees away from knowing 10 000 people. Three degrees gives you 100 times more people — or one million people. Four degrees gives you 100 million people, while five degrees gives you 10 billion people — almost twice the population of the planet.

Of course, the weakness in this argument is that while you might have 100 friends, and each of your friends might have 100 friends, they're not all _different_ friends. You will have many friends in

common. So this simple mathematical example is so simple, it's almost useless.

The Scientists Start Thinking

But in mid-1998, a couple of mathematicians developed a more sophisticated theory.

Duncan Watts (who grew up in Toowoomba, in southeast Queensland in Australia) and Steven Strogatz (who was born in Connecticut, in the northeast of the United States) were interested in networks.

Our lives are surrounded by networks: the fireflies that flash together at sunset; the internal circuits in the chips in your computer; the network of your friends; the financial network that starts with buyers and sellers and finishes with countries trading with each other; the network of nerves in your body; the network that we call the Internet, and so on. A network links a bunch of "things" together. In your average

You can get from Marilyn Monroe to Kevin Bacon in four links, or degrees.

Marilyn Monroe starred in *The Misfits* with Montgomery Clift (first link). Clift and Elizabeth Taylor worked together in *A Place in the Sun* (second link). Taylor and Roddy McDowall starred in *Lassie Come Home* (third link). McDowall worked with Bacon in *The Big Picture* (fourth link).

In another example, many people who grew up in Toowoomba (where scientist Duncan Watts grew up) are only one degree away from Watts, and two degrees away from Watts' sister, who is a diplomat. This puts them only three degrees away from the Cambodian leader Prince Norodom Ranariddh.

BACON CAN'T PLAY BACON

In a *Newsweek* interview, Kevin Bacon cheerfully admitted that he couldn't play the "Six Degrees of Kevin Bacon" game very well. The consolation is that he is the only actor with a game named after him.

network, some "things" are joined directly to each other, and some are not. So sometimes, to get from one place to another, you need a single jump, and sometimes you need a few jumps or links.

The mathematicians discovered a measurement that lets them get a grip on this fuzzy concept of "networks". It was the Average Path Length, i.e., the smallest number of links that it takes to join one point in a network to another.

Actors, Power Grid and Worms

The scientists looked at three separate networks: the Kevin Bacon Game (Collaboration Grid, if you want to get technical) in movie actors; the electrical power grid of the western United States; and the arrangement of the nerves of the worm *Caenorhabditis elegans*.

These were all well-known networks.

In April 1997, when the scientists looked, there were some 225 226 actors who were linked by the movies in which they had worked together. There were also 4941 generators, transformers and substations in the western United States electrical power grid joined by high-voltage power lines to make another kind of network. For many years, the neuroscientists have been looking at the 282 nerve cells in the little worm and have mapped out every single connection between the nerve cells in its nerve network.

RELATED TO THE PAST

If you go back about 30 generations into your past, you will find that you (and all your friends today) are related to just about everybody who lived in your particular area, or belonged to your particular culture. Bernard Derrida, from the École Normale Supérieure in Paris, and his colleagues discovered this when they mathematically analysed the family tree of Edward III of England (1312–1377).

Now people have always been worried by a paradox about the number of people in the past.

First, we humans have been around, as Modern Humans, for about 100 000 years. Back then, the total number of people on the whole planet was probably less than 10 000.

Second, try this calculation. You have two (2) parents, four (2 x 2) grandparents, eight (2 x 2 x 2) great-grandparents, and so on. As you go back another generation, you double your number of ancestors. A generation is about 25 years, so we Modern Humans have had about 4000 generations. So 100 000 years ago, surely you had a huge number of ancestors: 2 x 2 x 2 . . . multiplied some 4000 times, which is roughly equal to 1 followed by 1200 zeros. This is an extremely large number. A million has six zeroes, a billion has only nine zeroes, and the number of atoms in the Universe has about 80 zeroes in it.

How do you solve this paradox?

Easy. Practically everybody in the past appears in more than one family tree. For example, your grandparents might appear in 20 different family trees, if they had 20 grandchildren.

So, as you go back further in time, you are related to more of the people who were alive back then.

Two Types of Networks

Networks range from completely random to completely organised in arrangement.

People use regular, organised networks all the time. However, networks have a big disadvantage. It takes quite a few jumps to get from one point to another. Imagine a countryside organised into a regular, repeating grid. If you want to get from one side of the country to the other, you have to cross all the grid points in between.

On the other hand, in a totally random network, you can get from one point to another in a very small number of jumps. But a totally random network is almost useless because you can't be sure that you'll be able to do anything with it, or get to where you need. However, if you're lucky enough to stumble across a long link that goes most of the way across the country, you can travel across this random network very quickly.

MILGRAM STARTED IT . . .

Back in 1967 Stanley Milgram, a psychologist, published his paper called "The Small World Problem". He set up an experiment to measure just how many links it took to get a letter from one random American via a network of friends to another random American. Note that the links were to be via first-name friends, not business acquaintances, relatives or any other links.

   He did two separate studies. One starter-city was Omaha, Nebraska, while the other was Wichita, Kansas. In each study, the target-city was Cambridge in Massachusetts. For each study, he selected 150 random starter-people. They were each given a document folder with the following four items:

- The name and address of the target-person.
- The rules. One of the most important rules was, *"If you do not know the target-person on a first-name basis, then pass the document folder on to one friend that you feel is most likely to know the target. That friend must be someone you know on a first-name basis."*
- The roster. As each person received the document folder, they had to add their name to a list of names. This stopped the document folder from going back to somebody who had already received it.
- Tracer cards. As each person received the document folder, they had to remove a tracer card, fill it out, and mail it back to Milgram.

   Back in 1967, the results were astounding. On average, it took five friends to link one American to another random American. The number of links ranged from two to 10.

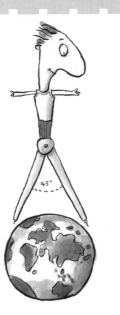

A Friend Makes All the Difference . . .

When Watts and Strogatz analysed these three real life networks, they were surprised by the results. They knew that these networks were *organised* enough to be useful, but they found that they were *random* enough to be fast.

   The Average Path Length is very small in a random network. The scientists worked out just what the Average Path Length should be in the three networks, if the networks were totally random. They then measured exactly what the Average Path Length was in real life. They were very surprised to find that the two measurements were really close.

How could this be?

They found that just a few random links make all the difference. A few random links can turn a large, remote, scary world into a small, cuddly, friendly world. (This is why scientists working in this field talk about the Small-World Network.)

Short cuts make the world smaller. In your workplace, they might speed up the flow of useful information — and nasty rumours as well.

Real Life Applications

There was a certain amount of "randomness" in the three networks the scientists looked at, so they guessed that it might turn out that many other networks in the real world also have a random structure.

For example, it turns out that in the network of nerve cells in the human brain, most of the nerves are linked to other nerve cells nearby. But occasionally they're linked to nerve cells way over on the other side of the brain, and nobody knows why. Perhaps these remote cells can fire up in the path of an expanding epileptic electric wave in your brain. In the same way that a firebreak can stop an expanding fire, these remote cells could contain an epileptic attack before it really takes off.

In the world of sexually transmitted diseases, most people have just a few partners, but a few have many. It takes just one or two very sexually active people to kick-start an epidemic. Some scientists are blaming the rapid spread of AIDS in India on a small number of promiscuous truck drivers. This is an unfortunate example of just how successful a Small-World Network can be.

And in your phone network, having one or two odd connections to unexpected places will make the overall phone network run more rapidly. Maybe it would be good to have a few randomly located towers for cellular phones. The same thing could work for the Net.

All sorts of people are looking at this research: economists, civil engineers, advertisers, sociologists, and health professionals.

People have always said that it's a small world, but understanding why it's small might pay off in a big way.

REFERENCES

Martha Claudia, Rollins Turner, David H. Kaplan, "Close Calls" (Letters), *Science News*, Vol. 154, No. 16, p 243.

James J. Collins and Carson C. Chow, "It's a Small World", *Nature*, Vol. 393, 4 June 1998, pp 409–410.

Paul Hoffman, "Man of Numbers", *Discover*, July 1998, pp 118–123.

Ivars Peterson, "Close Connections", *Science News*, Vol. 154, No. 8, p 124.

Duncan J. Watts and Steven H. Strogatz, "Collective Dynamics of 'Small-World' Networks", *Nature*, Vol. 393, 4 June 1998, pp 440–442.

INTERNET – NOT BUILT FOR THE MILITARY

The Internet (also called the Net) has grown with amazing speed. It seemed to come from nowhere and spread quickly across much of the world. Radio took 38 years to reach 50 million people, and TV took 13 years, but the Net took only four years. Unlike other media, the Internet is incredibly robust, and will keep on working even if most of it is destroyed.

The story goes that the Internet was designed and built as a military communications network that would survive a nuclear war. But that's not the whole story.

The Net — Intercourse, Not War

The Net was designed for the military as a robust communications system. But they didn't want it built, so it wasn't.

The next time the idea of the Net came up was in discussion about how to let computers talk to each other. The nerds needed to share expensive computer time between as many people as possible, so that, for example, a computer user in California could log onto an idle computer in Washington, D.C. More users made computer time cheaper.

When the Net first started up, computers were very rare, very expensive, and filled several air-conditioned rooms.

Hub-and-Spoke or Chicken Wire

Right from the very beginning, the Net was wired up differently from that other network that we all use — the phone network.

Your phone system has a hub-and-spoke arrangement. The hubs are the individual major switchboards around the country, while the spokes are the wires that lead out from these switchboards into your house. If you destroy enough of these hubs, or the big fat communications cables joining these hubs together, you've basically wrecked the phone system.

But the Internet is different. Think of the Internet as a giant mesh of chicken wire over an entire continent. There are several million strands of wire running east to west across the continent. To stop communications between the east and west sides of the continent, you have to break every single one of those strands, which is almost impossible.

Early Computers and the Cold War

The world's first digital electronic computer, ENIAC, fired up at the University of Pennsylvania in 1946. Over the next decade, the number of computers in the world increased very slowly. These computers were all very large and very expensive, and only a few countries had them.

Meanwhile, following World War II, relations between the United States and the Soviet Union became so chilly that we entered a period that became known as the Cold War. (The Cold War finished in 1991, when the then-Soviet Union could no longer pay the financial cost of matching the United States in military spending. The Soviet Union disintegrated.) Back then, defeating or at least containing the Soviets became the main priority of American strategic thinking. In 1946, in the United States, the National Security Act authorised the launch of the Central Intelligence Agency and the National Security Agency to help fight the "Soviet menace".

War Communications Wouldn't Work

The United States Air Force sponsored a thinktank called the RAND (Research and Development) Corporation in Santa Monica, California. During the late 1950s and early 1960s, much of the nuclear strategic thinking of the RAND Corporation became the official policy of the United States. Paul Baran joined the RAND Corporation in 1959.

At that time, the main deterrent against a Soviet nuclear attack was the threat of total nuclear retaliation by the Americans. But Paul Baran, like everyone else in the RAND Corporation, knew that the communication links between various elements of the United States nuclear retaliatory forces were dangerously weak and fragile. In a real attack and counterattack, many of the communication links probably wouldn't work.

The people at RAND had been trying to invent an impregnable communications system, but with no success. It was Paul Baran who came up with three radical concepts to make a communications system that would continue to function in any situation. These concepts ultimately led to the Internet.

The Breakthrough

The three concepts were a *distributed network*, *massive redundancy*, and *breaking up the message into many parts*.

A *distributed network* has advantages (and disadvantages) over a centralised one. If you destroy the single central control system of a centralised network, it stops working. But a distributed network can keep working, even after a lot of damage. A disadvantage of a distributed network is that you don't have as much control over it as you can have over a centralised network. This loss of control bothered the military.

The second concept was to have *massive amounts of redundancy* built into the network. So if a message couldn't get

through on one pathway, it might be able to get there through another pathway, say, via Japan and Africa.

The third idea was to *break up the message into many parts*. This was very radical. Quite independently, around the same time, Donald Watts Davies of the British National Physical Laboratory also came up with the same idea of breaking up digital messages into tiny bits that he called "packets".

Each packet would contain only a tiny part of the message, as well as information on where its final address was, and how to reassemble the packets in the right order. So on the Internet, your email message to your next-door neighbour might be broken up into hundreds of packets. Some packets might have gone directly via local telephone

links, while other packets might have gone via India or the United Kingdom, through underwater cable links, or even satellite, to get to their final destination — next door.

Too Revolutionary for the Military . . .

This was all very revolutionary, and only a handful of communications experts in the United States understood how clever it was.

WORLD WAR III ON THE NET — 1

Another myth is that critical military systems were connected to the Net in its early days. This is the premise behind the movie *War Games,* where a teenager stumbles into the main computer system in NORAD (North American Air Defense) and almost sets off World War III.

But NORAD, and other critical military systems, were never connected to the Net. Both NORAD and the Launch Control Headquarters (LCH) of the Strategic Air Command (SAC) in Omaha, Nebraska, were deliberately kept totally isolated from the Internet. So nobody could use the Net to get into NORAD or SAC.

NORAD is in charge of finding out what threatens the United States from air and space and passing on this information to the President of the United States.

NORAD is located just south of Colorado Springs, in Colorado. In the late 1960s, the military looked for a mountain that was as central as possible in the United States (to give the maximum warning time) and as strong as possible (to shrug off enemy nuclear weapon attack). They chose Cheyenne Mountain near Colorado Springs. As it turned out, there were some geological defects in Cheyenne Mountain, which they had to repair. They removed some 690 000 tonnes of granite to make a cavity about 400 metres inside the mountain — and a nice little parking lot outside. NORAD was designed to take direct hits from some 10 nuclear weapons and keep on working.

The so-called Big Red Button that sets off World War III is not in NORAD. It's always close to the President and carried by a high-ranking military officer. It's called "the football", because that officer is not allowed to let go of it or drop it. It contains the computer codes for that day that will unleash the full nuclear missile capability of the United States.

WORLD WAR III ON THE NET — 2

The President's Commission on Critical Infrastructure Protection was released in September of 1997. It defined the eight key elements of United States infrastructure as: electric power distribution; telecommunications; banking and finance; water; transportation; oil and gas storage and transportation; emergency services and government services.

Each of these infrastructures can today be "attacked" over the Net by "unfriendly" people. Today, strategic analysts talk about "cyberwar", "cyberterrorism", and "infowar". For example, over a trillion (a million billion) dollars are transferred every day over the Net. Disturbing that regular flow would create a catastrophe.

At the moment, the largest "machine" ever built by the human race is the telephone system. One day soon, it may be overtaken by the Net.

People in charge of military communications had no idea what Baran was talking about.

Paul Baran came up with the concept of the Internet as a way for the military to retain communication links that would survive a nuclear war. It was their bad luck that they could not understand what he was talking about, so the Internet was never connected to any strategic nuclear system.

In 1964, Baran gave up. "*So I told my friends in the Pentagon to abort this entire program — because they wouldn't get it right,*" he said.

But Perfect for ARPA

His brilliant idea languished until, in the late 1960s, Larry Roberts came across Baran's work for the RAND Corporation. Larry Roberts was an official in the Advanced Research Projects Agency (ARPA) of the United States Department of Defense. Roberts was not interested in a strategic military communications network. He wanted to get people across the United States to share the use and the cost of large, expensive computers. He realised that Baran's idea was exactly what he was looking for.

The theory was one thing, but actually building it was another. Nobody had ever built anything like this before, so they had to invent everything from the ground up. They had to work out how big each packet should be, how to track the packets, how to design, build and repair the special computers that would send the little packets to where they were supposed to go, and how to build all of the necessary hardware that would support the system.

Finally, on 1 October 1969, after a lot of hard work, two computers were ready to talk to each other. One was in Menlo Park in California, while the other was at the University of California in Los Angeles. The Los Angeles computer sent the word "LOG" to the Menlo Park computer, which immediately crashed.

At that exact moment, the ARPANET was born. It took another 14 years for the ARPANET to evolve into the Internet.

First, the ARPANET

Originally, ARPANET linked together defence contractors, universities and research laboratories. But soon, a strange mix of military people, anarchists, academics, science fiction fans, hackers, hippies, and people who just plain loved new technology had all jumped on board. Very quickly, they started using electronic mail. Soon, email became the most useful application (the "killer application" that everybody had to have) of the ARPANET.

The ARPANET had very few deep secrets; everything was "open" to see. Military strategists were used to believing that "important stuff" was secret, and so they thought that something as open as ARPANET was useless to them. So the military kept away.

As ARPANET evolved, the computers became faster and more powerful, and the rules that governed how packets moved changed. But around 1982–1983, there was a change back to a single protocol. TCP/IP (otherwise known as Transmission Control Protocol/Internet Protocol) was brought in to control the packets. This was a very important "invention". Because everybody could now follow the same standard (e.g.,

EMAIL

The word "email" can be spelt four ways.

"email" is very much the most popular spelling, followed by the less popular "e-mail". "E-mail" and "Email" are hardly ever used.

The opposite of email is snail-mail — your regular hand-delivered paper mail.

for labelling their packets), different small nets could talk to each other.

In 1983, the Defense Communications Agency split the network into two parts: ARPANET, basically for the universities and anybody else doing research; and MILNET, for nonclassified military communications.

Then the Internet

Also in 1983, ARPANET was officially renamed "the Internet".

Way back in the early days, every organisation connected to the ARPANET needed some kind of Department of

NETIQUETTE

This is a word made up from the words "network" and "etiquette". Polite Netizens (Net citizens) use netiquette. Netiquette (i.e., good manners) can take many forms.

For example, you generally don't use CAPITAL LETTERS in an email. Using CAPITAL LETTERS is called "shouting", and is considered rude.

Also, you shouldn't try to sell stuff to netizens by sending them unwanted email. This is called "spamming".

Defense (DoD) connection. But by the mid-1980s, organisations that had no link with the DoD started joining the ARPANET. By now a growing number of people were talking about the "Internet", or "the Net".

In 1986, the National Science Foundation (NSF) built its own "Net": the NSFnet. NSFnet linked the NSF's five supercomputer centres that were scattered over the United States. Gradually, NSFnet took over traffic from ARPANET.

By 1990, the NSF had closed down the original ARPANET "pipes" (communication networks). By 1994, all of the NSFnet had been sold to big telecommunications companies. The backbone of the Internet was now strictly commercial.

And it was that year, 1994, the Internet became available to the average citizen.

1994 — The Net Arrives

Until 1994, the major use of the Net was email. However, this soon changed dramatically.

chat line

TLA

Once you start using the **Net**, you run into lots of TLAs — Three Letter Acronyms. (However, just as the rude "four-letter words" don't all have four letters, not all the TLAs have just three letters.)

TDMTLA means Too Damn Many Three Letter Acronyms. However, the problem can't get too bad, if you stick to just three letters in your TLA. If you use the 26 letters of the English alphabet, there are only 17 576 (= 26 x 26 x 26) possible combinations of three letters.

Common TLAs include:

FAQ	Frequently Asked Questions
FYI	for your information
BTW	by the way
IMO	in my opinion
IMHO	in my humble opinion
IMNSHO	in my not so humble opinion
FWIW	for what it's worth
LOL	laughing out loud (lots of laughter)
TIA	thanks, in advance
NRN	no reply necessary
VR	virtual reality
RL	real life (opposite to VR)

Before 1994, nobody expected that the main future application of the Net would be the "browser". The browser would let you cruise, or browse, from place to place in the Net. It would browse across the World Wide Web (also called the "Web"). The browser ultimately became so popular that people called it the "killer application". Today, two very popular browsers are Netscape Navigator and Microsoft Explorer.

Multimedia

I remember being amazed that I could use my browser to download, several times per day, weather pictures of Australia. I would arrange these pictures into a little movie, and watch the clouds, storms and cold fronts rage across Australia. I would play this movie over and over again.

Back in 1990 and 1991, when I presented the weather on TV, the only way you could get pictures like this was to own a TV station, and have a big receiving dish on the roof or in the backyard. And here I was, getting them over a phone line! The multimedia potential of the Net really captured people's imagination.

Hypertext

The other thing that was fantastic about the Net was hypertext. In a document, a piece

SEEING THE FUTURE

J. C. R. Licklider, a brilliant psychologist and computer scientist, wrote about a time when *"human brains and computing machines will be coupled . . . tightly, and . . . the resulting partnership will think as no human brain has ever thought and process data in a way not approached by the information-handling machines we know today"*. When he wrote this, computers were huge machines without screens. All they did was "crunch numbers" (i.e., do addition and subtractions), and they certainly didn't "talk" to each other. Licklider was hired by Advanced Research Projects Agency in 1962.

of text might be highlighted. It wasn't ordinary "text"; it was "hypertext".

When you clicked on hypertext, a computer "connected" to that piece of text would start sending information to your computer. It didn't matter if that computer was in the next room or on the other side of the world, it would send information to you. A slab of, say, 1000 words might have 12 highlighted sections and could be connected to computers in 12 different countries.

The Net

The Net changed from something the military could not understand to something nobody could have imagined. Nobody had any idea that electronic mail would be one of the early popular uses of the Internet, or that multimedia and hypertext were the next big features that would capture the public imagination.

Who knows what the Internet, or World Wide Web (WWW) as it's also called today, will turn into in the next 10 or 20 years?

But there's one thing I would like to change. Why call it WWW? Why not save time and call it Six U?

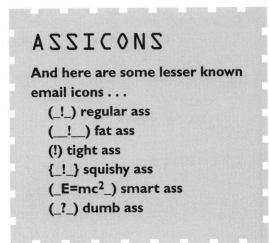

ASSICONS

And here are some lesser known email icons . . .

(_!_) regular ass
(__!__) fat ass
(!) tight ass
{_!_} squishy ass
(_E=mc^2_) smart ass
(_?_) dumb ass

INTERNET LOOKS FOR ALIENS

The Universe is a big place. The SETI (Search for Extra Terrestrial Intelligence) program looks for various electromagnetic signals that the aliens might send out. The trouble is that there are billions of frequencies the aliens might use, and there are billions of directions in the sky to search.

We need massive computing power to analyse all the signals that the SETI program captures. In mid-1999, the largest supercomputer you could buy had the equivalent of 9000 Pentiums. But what if you used all the computers that had access to the Net?

SETI@home is a project started by SETI scientists at the University of California at Berkeley. People can download a piece of free screen saver software to analyse the SETI data, and then download blocks of data. A typical block of data will represent 100 seconds of data from a tiny patch covering one millionth of the sky. You just let your computer analyse the data while you sleep.

In just the first fortnight, nearly 400 000 computer users signed up, bringing with them the equivalent computing power of 400 000 Pentiums. They donated some 3150 computer years of time to analysing the data . . . and found nothing.

If you look, you have a chance of finding aliens. But if you don't look, you definitely won't find them. And what if *your* computer found the first sign of aliens . . . ?

REFERENCES

Katie Hafner and Matthew Lyon, *Where Wizards Stay Up Late: The Origins of the Internet*, Simon & Schuster, New York, 1996.

Chris Taylor, "Monitoring the Aliens", *Time*, 7 June 1999, p 63.

102 Fidgeting Fat, Exploding Meat, & Gobbling Whirly Birds

EXPLOSIVES TENDERISE MEAT

We humans have been making explosives for 1000 years, ever since the Chinese invented gunpowder. Explosives have been used in many different ways, both in war and in peacetime. But just recently, a man who used to make nuclear weapons has come up with a new use for explosives — to tenderise meat.

Explosives

Basically, an explosive is something that very quickly generates a huge volume of gas. This gas then expands very rapidly, pushing anything around it out of the way.

An explosion is just a rapid burning. If you can make any substance burn rapidly enough, you have an explosion. So when a spark sets off a flame front that travels quickly through coal dust, fine sawdust, and even flour, soap, sugar or starch, it can produce a violent explosion.

The beautiful thing about explosives is that they store an enormous amount of energy in a tiny package. A big electrical power station covering a few square kilometres generates about 4,000 megawatts of power. A kilogram of dynamite,

smaller than a litre of milk, will generate 5000 megawatts of power — but only for a few thousandths of a second.

Air Pressure Does It All

Down at ground level, the average air pressure is about 10 tonnes per square metre (8.3 tons per square yard). This is one atmosphere of pressure. Imagine that you mark out a square on the ground — about 1 metre by 1 metre. Above that square is a column of air reaching up to space. The weight of all that air is about 10 tonnes.

Pressure is transmitted equally in all directions. There are 10 tonnes of air pushing against each square metre on the outside of the walls of your house. Why don't your walls fall over? Because there are 10 tonnes of air pushing against each square metre of the inside of your walls, which exactly balances the push from outside.

This is why pressure is so powerful. Suppose an explosion happens some distance away from your house. By the time the shock wave gets to your house, the pressure has dropped to 3 atmospheres (30 tonnes per square metre) above normal atmospheric pressure. Suppose that you have a 1 square metre window made of glass. The 30 tonnes of "push" will easily shatter your glass window.

Different Types of Explosives

There are mechanical explosives (such as a cylinder of compressed gas), nuclear explosives (either atom or hydrogen bombs), and the familiar chemical explosives (which can be solid, liquid or gas).

Mechanical explosives are used in mining, where you don't want to have huge amounts of gases generated. In underground mines, the excess gases can't escape easily, and can wreak havoc as they try to get out. The pressure from these

NOBEL PRIZE

Alfred Nobel was not just inventive, he was also a successful entrepreneur. He became enormously wealthy from his 355-plus patents.

Nobel was horrified by war, and he originally thought that his invention of dynamite would make war too terrible to wage. He was very wrong.

His will stipulated that a part of his enormous fortune should be used to provide a reward to be *"distributed annually in the form of prizes to those who during the preceding year had conferred the greatest benefit on mankind"*. Nobel Prizes were originally given in the areas of peace, physics, chemistry, physiology or medicine, and literature.

The prize for economics (which was added in 1969) is not a Nobel Prize. It is "The Bank of Sweden Prize in Economic Sciences in Memory of Alfred Nobel". The money comes from the Bank of Sweden.

However, there is no Nobel Prize for mathematics. Nobody really knows why. The rumour goes that there was an influential mathematician in the Academy whom Nobel didn't like.

The Nobel Prizes are awarded in the Concert House in Stockholm. After the ceremony, the guests are transported to the City House, for a Nobel dinner followed by Nobel ice-cream and Nobel entertainment. After that, the guests dance in a ballroom called the Blue Hall. The architect had planned it to be blue, but it was in fact decorated in a reddish colour. But it is still is called by its original name, the Blue Hall.

Nobel Prizes are awarded on the day Nobel died, 10 December. Scientists craving the Nobel Prize talk about "going to Stockholm" or "going to the Blue Hall".

gases can knock over supports, or bring down walls.

The most powerful explosives are nuclear explosives — but they don't have many peaceful uses.

The most commonly used explosives are chemical explosives. A chemical explosive can be either solid, liquid or gas. In general, liquid and solid chemical explosives will give you a more powerful explosion than a gas explosive.

"High" and "Low" Explosives

Most chemical explosives have nitrogen atoms in them. When the explosion happens, oxygen molecules break loose from the nitrogen and form new chemicals with carbon and hydrogen. You end up with carbon dioxide, carbon monoxide, water vapour, nitrogen molecules — and a huge amount of energy. From 1 g of explosive, you'll usually get about 1000 Cal of energy and about one litre of gas.

In chemical explosives, you have two choices: low explosives, or high explosives.

In a low explosive, the burning flame makes the explosive go "bang". In a high explosive, an internal shock wave sets off the "bang".

Low Explosives

A low explosive detonates slowly. It takes a few thousandths of a second to explode. There is a "burning front", which burns through the low explosive slower than the speed of sound in that material — between 0.03 and 350 kph (0.02–220 mph). As each "section" burns, it gives off gas and heat. The heat makes the next "section" of low explosive burn.

EXPLODING HORSES

Suppose that you are a Ranger in a United States National Park, and that you have just stumbled across a large dead horse. The danger is that a bear might be attracted to the horse and then, after a little nibble, start looking for humans to eat.

What is the official recommendation of the United States Department of Agriculture Forestry Service?

Jim Tour gave the answer in the January 1995 issue of *Tech Tips*, published by Technology & Development Program, and entitled "Obliterating Animal Carcasses with Explosives".

Yep, the answer is explosives. If necessary, use up to 25 kg (55 lb) *"simply draped over the carcass"*.

The *Tech Tips* really do live up to their name. Very wisely, they advise that *"Horseshoes should be removed to minimise dangerous flying debris"*.

Low explosives are often used in firearms, where the expanding gases push a projectile out of a barrel, and in fireworks. Gunpowder (black powder) is a typical low explosive.

High Explosives

A high explosive detonates very quickly, because its molecules are very unstable. In a high explosive, the detonation happens in a few millionths of a second. When you set it off, a shock wave zips through the high explosive faster than the speed of sound through that material — between 3500 and 35 000 kph (2220–22 200 mph). As the shock wave passes, it leaves behind a trail of broken-down chemicals. These chemicals, as they break down, give off gas.

A powerful high explosive can generate pressures as high as one quarter of a million atmospheres. A quarter of a million atmospheres means that 2 500 000 tonnes are pushing on each square metre! You can see why a high explosive can push a building over. If you tried to use a high explosive in a gun, the barrel would simply shatter from the incredibly rapid increase in pressure. Typical high explosives include TNT (trinitrotoluene, $C_7H_5O_6N_3$) and dynamite, which are used in military weapons and in mining.

There are two types of high explosives: primary and secondary.

Primary High Explosives

You can easily set off a primary high explosive with any reasonable amount of heat (such as friction, a spark, a flame, or an impact that produces heat). Lead azide and mercury fulminate are primary high explosives.

Secondary High Explosives

But a secondary high explosive needs a detonator. You can't fire it off with a match — that will only make it burn relatively harmlessly, giving off huge amounts of gas. But if you set off some mercury fulminate next to a high explosive, it will detonate with a massive shock wave.

History of Explosives

The very first explosives were just lengths of bamboo. Bamboo is hollow, and divided into sealed sections. Around the 2nd century BC, short sections of bamboo were thrown into fires as a simple firework for celebrations and festivals. The internal air expanded and blew the bamboo open with a loud noise — making the noise "Ba-boom", and giving the name "Bamboo".

GREAT INVENTIONS IN EXPLOSIVES

One major problem with explosives is being too close to them when they go "bang".

In 1831, William Bickford from England invented the safety fuse. It burnt at a constant rate, so you could time the explosion by the length of safety fuse you laid out. The first version was just a narrow tube of gunpowder wrapped in a textile fabric, such as jute.

When nitroglycerin was invented, one problem with it was how to explode it reliably and safely. Nobel invented a blasting cap in 1865 that did just this.

Towards the end of the 19th century, engineers began to use electrical firing, perfected by the American H. Julius Smith. This allowed them to time explosions more accurately.

Explosives have been around for 1000 years, since the Chinese invented gunpowder. Gunpowder is about 75% saltpetre (potassium nitrate), 15% charcoal, and about 10% sulphur. In the early days, the saltpetre came from animal wastes, or compost piles.

The Arabs knew of gunpowder by the 1200s. In 1242, the English philosopher Roger Bacon wrote a heavily encoded document in Latin, describing how to make gunpowder. Around 1304, the Arabs invented the first real gun. It was a bamboo pipe reinforced with iron that fired an arrow. This simple invention changed the face of Europe.

A cannon was just a bigger gun. Once cannons came onto the scene, knights were no longer safe in their castles, and the feudal system began to crumble. By 1314, we have a written record of guns and gunpowder being shipped from Belgium to England. This was high-tech stuff, and the arms race was well under way.

At first, gunpowder was used only by the military.

Peaceful Explosives

Explosives were first used peacefully in civil engineering to make holes in the ground. One of the earliest peaceful use of explosives was in 1679, in the making of the Malpas Tunnel of the Canal Du Midi in Languedoc, France. Soon after, the engineers began to appreciate the incredible power locked up in explosives, and began to use them to mine and quarry, and to build roads, canals, harbours and underground railways.

But black powder was only a low explosive. The engineers needed more blasting power, so the chemists gave them high explosives.

Nitroglycerin

Ascanio Sobrero, an Italian professor of chemistry in Torino, invented nitroglycerin ($C_3H_5N_3O_9$) in 1846. Nitroglycerin is a very poisonous pale yellow oil that is also very unstable. The molecules will split even if the oil is just shaken or heated. It freezes around 11°C (52°F), so you can easily freeze it just by packing it in ice at 0°C (32°F). Frozen

NITROGLYCERIN, THE DRUG

Nitroglycerin is not just a high explosive. It's also a medical drug that can treat the pain of heart disease.

Nitroglycerin makes blood vessels bigger. When somebody feels "angina" (the pain of heart disease), they can take a tablet that opens up the blood vessels around the heart.

Even a small increase in diameter of these blood vessels gives a big increase in flow. The flow of a liquid through a pipe increases as the fourth power of the diameter. In plain English, if the diameter of a blood vessel doubles, the blood flow increases by the fourth power of 2 — which is 16.

With this extra blood flow, the muscles of the heart get enough oxygen, and the angina disappears.

OTHER EXPLOSIVES

Chemists have been inventing explosives for a long time.

Salts of chloric or perchloric acid have been used to make explosives since 1788. Many factories were built in the United States and Europe to make these explosives, but they all eventually blew up, or burnt down. These types of explosives are no longer used.

Trinitrotoluene (TNT) is a high explosive, widely used by the military. It was invented in 1863, and it was originally used in the dye industry. It was first used as an explosive in 1904. One property that makes it very valuable to the military is that it can be melted and poured into various shapes, either on its own or combined with other explosives.

TNT was first used militarily in the Russo–Japanese War in 1905. Since then, TNT and its derivatives have become the most commonly used conventional military explosives.

In 1871, Hermann Sprengel in England combined various oxidising agents (nitrates, nitric acid or chlorates) with inflammable substances (benzene, nitrobenzene and nitronaphthalene). One oddity about these Sprengel explosives is that they were made by combining the ingredients immediately before they were to be used. Another oddity was that one of the ingredients was a liquid.

In 1876, Nobel received a patent for blasting gelatin, which he made from nitroglycerin and guncotton. Blasting gelatin was plastic enough to be moulded into various shapes, had impressive resistance to water, and was more powerful than dynamite.

In 1887, Nobel invented another explosive called "Ballistite". This smokeless powder was made of 60% nitroglycerin and 40% nitrocellulose. This ballistite was a superb propellant for firearms. The British ignored Nobel's patent and made their own version, which they called "Cordite".

In 1895, Carl von Linde in Germany invented a new explosive — carbon black, which had been packed into porous bags and then soaked in liquid oxygen. Like Sprengel's explosives, one of the ingredients was a liquid. This explosive became very popular in Germany in World War I, because of the difficulty of getting nitrates. Liquid oxygen explosives reached their maximum production around 1953 (about 10 000 tonnes per year), but were no longer used after 1968.

This happened because of a revolutionary new explosive developed in 1955 called ammonium nitrate fuel oil. Even though Linde explosives were cheap, ammonium nitrate fuel oil was even cheaper. Today, explosives based on ammonium nitrate fuel oil or ammonium nitrate water gels make up about 70% of all the high explosives used in the United States.

nitroglycerin is more stable and less dangerous. Sobrero had his face badly scarred by a nitroglycerin explosion. In fact, he thought nitroglycerin was much too dangerous ever to be widely used. He said, *"when I think of all the victims killed during nitroglycerin explosions, and the terrible havoc that has been wreaked which in all probability will continue to occur in the future, I am almost ashamed to admit to be its discoverer"*.

Alfred Nobel tamed nitroglycerin.

In the early 1860s, he and his father set up a small factory in Sweden to make nitroglycerin. In 1864, five people including his youngest brother Emil were killed in an enormous nitroglycerin explosion. The Swedish government thought that nitroglycerin was so unsafe that it would not let him rebuild his factory. He got around this by making nitroglycerin on a barge moored in the middle of a lake.

Other governments were also worried by nitroglycerin. In 1866, after ships had exploded in various harbours all around the world, many other nations made it illegal for their ships to carry nitroglycerin.

Dynamite

The word dynamite comes from the Greek *dynamis*, meaning "power". Nobel made

the breakthrough in 1867, when he patented what he called Dynamite No. 1 — 75% nitroglycerin and 25% kieselguhr. Kieselguhr, also called guhr, is a porous absorbent clay. The guhr absorbed the nitroglycerin, making it safer to handle.

Nobel soon realised that while the guhr made nitroglycerin safer, it did not contribute anything to the explosion. Instead, it absorbed heat and actually weakened the explosion. He found other ingredients such as wood pulp and sodium nitrate that would absorb the nitroglycerin and make it safer, yet give a more powerful explosion.

Dynamite was a great success and started Nobel on the road to fabulous wealth. Sobrero was a little upset, as he had done so much of the groundwork and received so little, compared to Nobel. But Nobel had made nitroglycerin safe, and he always acknowledged the work of Sobrero.

One spectacular use of dynamite occurred in 1885 when Flood Rock, which was seen as a *"menace to navigation"*, was excavated in New York Harbour. In a very impressive explosion, some 34 tonnes (33.5 tons) of No. 1 dynamite and 110 tonnes (108 tons) of potassium chlorate-nitrobenzene removed the danger.

C-4

If you watch a lot of action movies, you'll soon see somebody stuffing some "C–4" explosive in a narrow gap of a safe or door. C–4 has had a lot of free publicity. Terrorists love C–4 because it's easy to mould, can be set off electrically, and has one the highest energy yields of any non-nuclear explosive.

C–4 (and its relatives, C–1, C–2 and C–3) contains about 80% RDX (cyclotrimethylenetrinitramine) and 20% plasticisers, oils and waxes, to make it soft.

The "C" stands for "Composition". The different C-numbers relate to the different temperatures over which these explosives are soft and pliable, and do not leak or exude oil. So C–4 stays soft down to –57°C (–70°F) and will exude oil only above 77°C (170°F). C–3 is usable over a narrower range; it will stay soft down to –29°C (–20°F), and will weep oil only above 49°C (120°F).

TAGGING EXPLOSIVES

"Taggants" are tiny multi-coloured chips of plastic. They are all very different from each other. If they are added to an explosive, they will uniquely identify that explosive down to which factory made it, which batch it was in, on which date, who distributed it, and which shops sold it. So taggants can be used to track down improper use of explosives, such as by terrorists.

The most widely used taggants were invented back in the 1970s by Richard G. Livesay, when he worked for the 3M company. To the naked eye, the taggant looks like specks of black pepper. But under the microscope, you can see that they are made of up to 10 mini-slabs of coloured melamine plastic, bonded to a magnetic material. The different colours make up a unique bar code. After an explosion, the investigators can gather the taggants out of the debris with a magnet, look at them under a microscope, and recognise and identify the tag.

Livesay bought back the licence for his invention from 3M and set up his own company, Microtrace, in 1985. His company currently has only one client — the Swiss government. In Switzerland, the government has tagged both low and high explosives. It has been a sensible decision. Since 1984, the Swiss have solved some 560 cases of bombing by using these taggants.

In the early days of taggants, explosives manufacturers in the United States liked the idea. But today it seems as though they don't want taggants added to their products, in case they were to get sued for the damage caused by their explosives.

The conservative NRA (National Rifle Association) has also moved against taggants being added to gunpowder. They say that it's possible these taggants could make the gunpowder dangerous. The NRA claims that certain tests where taggants were added to gunpowder showed two bad effects. First, the gunpowder could become unstable, so that it might spontaneously explode. Second, the gunpowder might degrade more quickly than normal.

However, these tests were done with the taggants being added to the gunpowder at concentrations of 500 000 parts per million. The official concentration recommended by Microtrace is 250 parts per million — that's 2000 times less.

The Swiss have never had any problems with adding taggants to gunpowder and other explosives. And in the United States, between 1977 and 1980, three major explosives manufacturers (Dupont, Hercules and Atlas Powder Co.) successfully added taggants to about 3200 tonnes (3100 tons) of explosives without a single problem.

There has been a bit of a technology update with taggants — isotopes. Think back to school science. An atom has a central core (the nucleus) with a bunch of electrons rotating around it. The core has two types of particles in it — the protons and neutrons. An atom of hydrogen has one proton, an atom of carbon has six protons, and so on. In any element, the number of protons is always the same, but the number of neutrons can vary. An isotope of, say, carbon just has a different number of neutrons, but is otherwise chemically identical to the other isotopes of carbon.

Many atoms, such as carbon or nitrogen, have a few isotopes. Common laboratory techniques, such as gas chromatography and mass spectrometry, can easily pick up these almost invisible isotopes in concentrations as low as a few parts per billion.

The bomb in the Oklahoma City explosion was made from ammonium nitrate (basically, fertiliser) and fuel oil. In 1995, a company working with isotope tags showed that they would survive explosions of ammonium nitrate and fuel oil, and still be detectable. But soon the fertiliser companies followed the lead of the explosives companies. They moved away from using taggants, to trying to stop taggants being introduced.

With all the legal issues involved, and with terrorism still with us, tagging is quite an explosive topic.

More Uses for Explosives

You can use explosives to make perfectly smooth cuts through rock, or to weld together sheets of metals that are otherwise impossible to weld, such as steel and aluminium, or steel and titanium. The Nitro Nobel company of Sweden can weld a slab of stainless steel 12 mm (½ in.) thick onto any other steel, with slabs 6 metres by 2.5 metres (20 ft by 8.2 ft)! It is impossible to weld such large slabs together using any other technique.

Explosive welding is also used to make American coins. Large slabs of various metals are placed parallel to each other, about 6 mm apart. A special explosive is placed on top of the upper slab, and it rams the two slabs together so forcefully that they become welded. This new single slab is then squashed between two rollers to the desired thickness, and then coins are punched out of it.

Explosives can also be used to split a thin sheet of metal to exactly half the original thickness. It needs careful planning, though. The explosive is set off while it touches one side of the metal. A compression wave rushes through the metal to the other side, bounces off, and becomes a tension wave. Wherever the tension in the wave front is greater than the strength of the metal, the metal splits apart.

Explosives form metals into odd shapes that cannot be achieved by any other method. The first patent for this was taken out by an English engineer called Walter Claude Johnson, in 1897. It described "*a method of generating fluid pressure for tube jointing, or embossing, stamping of metal sheet and such like operations*".

Explosions are also used to "implode" buildings. A demolition that would take months of noisy and dusty work is shortened into weeks of preparation and a few seconds of collapse. A seven-storey building can be reduced to a 6-metre (20-ft) pile of rubble.

We use explosions all the time. Explosions of the fuel–air mixture inside the cylinders of your vehicle make the pistons go up and down.

Explosives Tenderise Meat . . .

But tenderising meat is a very unusual use for explosives. The old and expensive way to tenderise meat was to hang it in a cold store for a few weeks. It took a very unusual man to think up a brand new way to tenderise meat.

John B. Long has had a long and interesting career in science — and he hasn't slowed down. On 3 November 1998, he celebrated his 79th birthday. During World War II, he was a meteorologist. In the late 1940s he made robots to handle radioactive metals. At the time, he called them "remote-controlled equipment". Glenn T. Seaborg (after whom seaborgium, element 106, is named) used these robots to help discover 10 radioactive elements, including plutonium.

In the late 1950s, Long moved to the Lawrence Livermore National Laboratory, in San Francisco. He spent the rest of his career as a government scientist designing nuclear weapons.

And now in his new career as a self-employed scientist, Long is working out how to use explosives to tenderise meat in a fraction of a second.

NUCLEAR DEVICES

Nuclear weapons are hardly ever called nuclear weapons. They are usually called nuclear "devices".

Are they extremely powerful consumer devices: super can-openers, turbo dishwashers, or supersonic vacuum cleaners? (There were a few unsuccessful attempts to use nuclear weapons to make harbours.)

No, they are weapons. Their only use is to explode. But only two nuclear weapons were exploded in anger — over Hiroshima and Nagasaki, at the end of World War II.

At the height of the Cold War (1946–1991), there were some 50 000 nuclear weapons on the planet. Both the United States and the Soviet Union used them as a threat against each other. The policy was called MAD — Mutually Assured Destruction. If one side was to attack, the other side would respond with all of their nuclear weapons. In other words, the slightest attack would lead to complete and utter destruction of the aggressor as well as the attacked, and many other people who didn't belong to either the United States or the Soviet Union.

HOW NUKES WORK

Nuclear weapons rely on Einstein's famous equation:

$$E = Mc^2$$

where "E" is the energy, "M" is the mass, and "c" is the speed of light. My children and I actually saw the original equation in Einstein's own handwriting, when it came to Sydney in the late 1990s, as part of the publicity tour leading up to its sale.

"c" is a very large number. This means that a small amount of mass can be turned into a huge amount of energy.

In an "atom" or "fission" bomb, heavy atoms such as uranium or plutonium are broken into smaller atoms. In these nuclear reactions, some of the mass vanishes, but reappears as a huge amount of energy.

In a "hydrogen" or "fusion" bomb, light atoms such as hydrogen or lithium are "fused" together to make heavier atoms. Again, there is a small shortfall in mass, which appears as a huge amount of energy.

"Explosive Tenderisation" will save money and be good for the environment. Long calls the process of using explosives to tenderise meat the "Hydrodyne Process". The word Hydrodyne comes from *hydro* meaning water, and *dyne* from the Greek word meaning "power".

Swimming Pool Dreaming . . .

It all began in the late 1960s, while John Long was floating in his backyard swimming pool. At this stage in his career, he was designing triggering mechanisms for nuclear bombs. To be specific, he was working as an explosives expert (a type of mechanical engineer). As part of his experiments, his team detonated small chemical explosives underwater. He was worried what would happen if one of these explosives detonated accidentally, while the technicians were in the water installing them. After all, a human body has roughly the same density as water, so shock waves would travel easily through it, instead of being reflected. Depending on its size, the explosion could certainly do some unwanted damage, and even kill.

And then he started wondering what these shock waves would do to a piece of steak . . .

First, Explode Your Meat

Long's friends helped him do an experiment at a privately owned explosives testing site. They started with a slice of tough beef. They cut it in two — one part to be "exploded" or "shocked", the other to be left untouched (for comparison). They wrapped one slab of meat in plastic, filled a 55-US gallon (208-litre) paperboard

BIG CONVENTIONAL EXPLOSIONS

The biggest non-nuclear, commercial explosion in North America was set off on 5 April 1958. Ripple Rock, an underwater mountain with two peaks, lay just under the surface (2.7 metres or 9 ft) at low tide in the Seymour Narrows, between the British Columbian mainland and Vancouver Island. Over the years, more than 120 vessels had foundered on this twin-peaked submerged mountain.

Starting on the shore, a special tunnel was driven down to below the level of the sea floor, across to the underwater peaks, and up inside them. The end of the tunnel was packed with 1253 tonnes (1233 tons) of the powerful high explosive Nitramex. The enormous blast removed the top of this mountain, so that now it was a safe 15 metres (50 ft) under the surface at low tide.

But the biggest known conventional explosion was on 28 December 1992. Chinese army engineers used about 13 670 tonnes (13 450 tons) of TNT to obliterate a mountain in Zhuhai, near Macau. Once the mountain had gone, the local airport was expanded.

MILITARY EXPLOSIVES

Military explosives have different requirements from commercial explosives. They need a long storage life, without degrading. They also have to be resistant to accidental detonation, such as from bullets. Some have to be resistant to water damage.

In general, military explosives are rigid solids, while commercial explosives are soft and pliable so they can be stuffed into irregular holes.

MOST POWERFUL CHEMICAL EXPLOSIVE

Air Force scientists at Edwards Air Force Base in California have recently created the most powerful chemical explosive yet. Not all the details have been released, but it is known that the chemical is related to TNT. It is also known that the basic molecule is made from five nitrogen atoms linked to arsenic and fluorine atoms.

When it was being analysed, a few grains of the white crystalline powder were left behind in the analysis equipment. They were so powerful that when they accidentally exploded, they destroyed the equipment.

The chemists are trying to make this molecule even more powerful, by making it with 10 nitrogen atoms.

drum with water, and dropped the plastic-wrapped beef into the water. It sank to the bottom of the drum. Then they hung some C–4 chemical explosives in the water, not touching the meat. They retreated to a nearby bunker, where they could watch the action on a TV monitor, and set off the explosives. He said that the *"drum totally disappeared. There were just little pieces of paper fiber all over."* Eventually, after searching for 15 minutes, they found the meat on the side of a nearby hill.

John Long then cooked both the shocked and unshocked meats on a BBQ that they had brought along just for the occasion. He said that the unshocked meat was *"so tough you could hardly chew it. But the one we shocked — it was delightful, as tender as a $10 steak in those days."*

Today, butchers deal mostly with boneless meat. But back then, butchers worked mainly with whole sides of meat. Unfortunately, the bones in the meat interfered with the shock waves from the explosion. So the meat was uneven, with some parts tender, and others tough. At the time, they couldn't solve this problem.

John Long really didn't know what to do with his invention, so he put it aside for 20 years.

In 1988, he "retired" and moved to Florida. He soon teamed up with Stanford Klapper to make and sell inexpensive solar cookers. But once Klapper heard Long's musings about how explosives could tenderise meat, he suggested that they leave solar cookers, and instead explode some meat.

Tenderness of Meat

Tenderness is one of the most important qualities of meat for the meat eater. There has been a huge amount of research into tenderness over the last 50 years.

Meat varies in tenderness from the very tender (e.g., psoas major), up to the very tough (e.g., superficial pectoral muscle). There is actually a scientific method used to classify how tender meat is. You cook some meat, and then remove a 12-mm (½ in.) core, parallel to the muscle fibres. You then rest a sharp guillotine blade on the core. The weight needed for the blade to slice

EARLY EXPLOSIVE TENDERISATION

Unscrupulous fishers have been tossing explosives in the water for decades. But this was mainly to kill or stun the fish, so that it would be easy to catch them, not to tenderise them.

In 1970, a man called Godfrey also thought of using explosives to tenderise meat. He received a patent, which described wrapping the meat in a protective layer to exclude the air, before a supersonic shock wave would tenderise the meat. But he had troubles with some of the more important details, such as the relative positions of the meat, tank and explosives.

through the core of cooked meat ranges between 3 kg (6.6 lb, rated very tender) and 15 kg (33 lb, extremely leathery). In general, anything that is rated at less than 4.6 kg (10.1 lb) is thought of as tender.

Meat that is "marbled" with fat interwoven through the lean muscle is very tender and juicy, but we know that for some people, too much fat in their diet is a health risk. You can "dry-age" meat by hanging top-quality cuts in a well-controlled refrigerated environment, but this makes the meat very expensive, because of the cost of the storage and the fact that the meat actually loses a little weight during the process.

There are many different ways to make the meat more tender. You can chill the meat to various timetables, hang it for various lengths of time at various temperatures, give it electrical impulses, inject enzymes into the live animal or sprinkle the enzymes onto the carcass after it's dead, or try the good old-fashioned method of mechanical damage with multiple needles or blades. But these all take lots of time, energy and money.

That's why a quicker and cheaper way to tenderise meat was so attractive to Long and Klapper.

The Team Assembles . . .

So John Long set up the Hydrodyne company with Stanford Klapper. In 1992, Long and Klapper tried to convince Morse B. Solomon (head of Meat Science Studies at the United States Department of Agriculture's Beltsville Agricultural Research Center) about the potential of the "Hydrodyne" process.

The first Explosive Tenderisation that Long and Klapper demonstrated to Solomon was a total failure, but the following tests were extremely successful. They were good enough to convince Solomon to bring his agency in as a partner with Long and Klapper in the Hydrodyne company. Soon they brought Eric Staton into the team. He was trained and licensed to use explosives.

Their first detonation chambers were $US6 plastic garbage cans. They dug a hole in the ground, put a garbage can in the hole, filled it with water, put an 18-mm thick (¾-in.) metal disc on the bottom of the garbage can to act as a shock wave reflector, and added the meat and just a touch of C–4 explosive (later nitromethane and ammonium nitrate were used). The plastic

MICROSTRUCTURE OF MEAT

If you look at a muscle with the naked eye, you can see the individual muscle fibres. These muscle fibres are made up of many tiny myofibrils, which are in turn made up of thousands of tiny myofilaments. Each myofilament is made up of long skinny filaments called actin filaments and myosin filaments. They are each made up of many actin molecules (about 6–8 nm or 0.23–0.26 millionths of an inch in diameter), and myosin molecules (about 15 nm or 0.59 millionths of an inch in diameter).

The smallest unit inside a myofilament that can contract or relax is a collection of actin and myosin filaments called a sarcomere. The actin and myosin filaments are laid out so that they overlap each other. They have a strange, but very effective, motion where they "ratchet" over each other to change length.

The length of a sarcomere varies from 1–4 microns (40–160 millionths of an inch) depending on whether it's very contracted or very stretched.

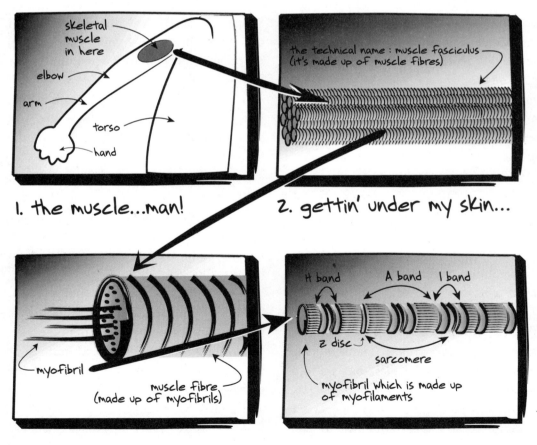

1. the muscle...man!

2. gettin' under my skin...

3. a little closer...

4. the fine print!

garbage cans worked beautifully, even though they didn't look very pretty, and were completely destroyed during each Explosive Tenderisation process.

The team spent about two years using the plastic garbage cans to work out the finer details — for example, how much explosive to use, and exactly where to place it in the container. They found that about 100 g (4 oz) of explosive approximately 30 cm (1 ft) from the meat would generate about 1700 atmospheres of pressure (17 000 tonnes per square metre, or 25 000 psi). This pressure would successfully tenderise chicken breasts, but other meats needed only half as much.

Solomon was impressed with the improved tenderness. He said, "*I can bring meat that starts with a shear force of 6, 8, or 12 kg down to 3 or 4 every time — sometimes lower*".

Hi-Tech Tenderisation

The disposable garbage cans worked well, but they were messy. And it was wasteful to destroy a new can every time.

It was time to go hi-tech with reusable cans. The team soon received an Energy-Related Invention Grant from the United States Department of Energy and the National Institute of Standards and Technology.

Around 1996, they designed and built a stainless steel tank 1.2 metres (4 ft) in diameter, which was inside a frame that absorbed the shock. The tank weighed some 3175 kg (7000 lb). The lid had a very pronounced dome (to contain the pressure) and weighed about 2270 kg (5000 lb). The tank was buried approximately 3 metres (10 ft) deep in the ground. The 1060-litre

(280-US gallon) tank could simultaneously tenderise 30 cuts of beef, each weighing about 9 kg (20 lb) — that's 270 kg (600 lb) in total.

The team realised that the meat had to be specially wrapped so that it wouldn't absorb water, or contaminants in the water. After all, a jet of high pressure air from a compressed-air hose can drive dirt into your skin and give you an instant unwanted tattoo.

In their technique, the meat is put in a polyolefin resin bag, which is then encapsulated in a rubber isoprene bag. All the air is sucked out of each bag. The meat is then placed at the bottom of the round-bottomed stainless steel tank of water. A small amount of explosive (50–100 g, or 2–4 oz) is set off some 30–60 cm (1–2 ft) away from the meat. It typically generates a shock wave with pressures of some 680 atmospheres (6800 tonnes per square metre, or 10 000 psi). But once one shock wave bounces off the steel walls, it can add to other shock waves, and generate pressures up to four times greater. This shock wave rips through the water and the meat (which is about 75% water).

In an explosion, the shock waves bounce off the metal sides of the container and re-enter the meat. In fact, the pressure is so high that the water gets pushed up into the lid, which is why it needs to be so heavy and in the shape of a dome.

Morse Solomon says that the Hydrodyne technique can increase the tenderness of certain cuts of beef by 72%. He said, "*The Hydrodyne process could be used by companies that sell meat to hotels, restaurants, and supermarkets. It could give top-quality tenderness to lower grade cuts.*"

FACTORS OF TENDERNESS

The main factors that affect the tenderness of meat are the *Actomyosin Effect*, the *Background Effect*, and the *Bulk Density* or *Lubrication Effect*.

In the *Actomyosin Effect*, meat is more tender if it has a longer sarcomere, a smaller muscle fibre diameter, and fewer sarcomeres per fragment of myofibril.

In the *Background Effect*, meat is more tender if it has fewer connective tissue proteins, if the elastin (an elastic protein) fibrils are smaller, and if the collagen (a protein in bone and cartilage, and which also makes up the white fibres in connective tissue) is more soluble.

In the *Bulk Density* or *Lubrication Effect*, meat is more tender if it has more marbling (or fat), and if that marbling is distributed widely and evenly throughout the meat, instead of being collected into small lumps.

MORE FACTORS OF TENDERNESS

A few other factors that affect the tenderness of meat are the breed of the animal, how much fat it has, how much work it has done, its age, and the degree of cooking.

Bos indicus breeds of cattle (Sahiwal, Brahman, etc.) tend to give a tougher meat than *Bos taurus* breeds (Hereford, Angus, etc.). After the animal dies, the muscles degrade and get more tender. But *Bos indicus* breeds have more of a protein called calpastatin that interferes with this muscle degradation, so the muscles stay harder.

One "advantage" of fat marbling is that it tenderises the meat as it cooks. The fat melts, spreads throughout the meat, and lubricates it.

As you exercise a muscle and make it stronger, it lays down more myofibrils. There are more proteins, so the meat is less tender.

As an animal gets older, the collagen in the connective tissue gets stronger and more rigid, making the meat tougher.

Another factor is the degree of "doneness". This is a technical term for how cooked it is. Meat gets tougher as it gets more "done" or cooked.

Why Explosions Tenderise Meat

When you examine the shocked meat with the electron microscope (see illustration on page 118), you can see that the shock wave tears the myofibrils into smaller fragments and breaks the protein in the cell walls of the meat, making the meat more tender. The shock wave tends to tear the myofibrils apart consistently in the same area — at the ends of the sarcomeres.

But the fats and oils that give flavour to meat are relatively unaffected by the shock. Another advantage is that the shock wave kills some of the bacteria that can spoil meat. Perhaps it ruptures the cell walls of bacteria. The shock wave also seems to kill the parasites in pork that cause trichinosis. (Trichinosis is an infection by a roundworm that migrates from your gut to your muscles. You can catch the roundworm by eating under-cooked meat, such as pork.)

Solomon loves shocking meat. He says that *this technology hasn't met a piece of meat it doesn't like*.

One Step Forward . . .

The high-tech stainless steel metal tank seemed like a great idea, but it doesn't work as well as the cheap $6 throw-away plastic garbage cans. First, the results are not as consistent, and vary from batch to batch. Second, the hi-tech reusable metal tank doesn't tenderise the meat as well as the old plastic can that disintegrated.

They're still trying to get around Murphy's Law. It's not clear exactly what the problem is, or how to fix it.

In the old plastic drum, the explosive would expand, and everything else (water and meat) would try to get away from the shock wave. In the new metal drum, the shock waves are trapped inside the tough drum. They get more than one pass at the meat. Perhaps the number of passes depends on the exact relative positions of the meat, explosive and drum, and any tiny change in position leads to a big change in shock wave strength.

They may even have to abandon the $US1.2 million recyclable drum and go back to $US6 disposable drums. This might break their hearts, after they have invested $US1.2 million and many working years in the stainless steel tank.

Advantages of Explosive Tenderisation

Each year in the United States, some 37 million head of cattle are slaughtered for the meat market. While this meat ages in the refrigerator to get more tender, it consumes about $US14 million worth of electricity per year — about 320 million kilowatt-hours. It takes about two weeks to get this meat to market. If they could make the meat more tender in a fraction of a second, this time could be reduced to a few days, resulting in huge savings to power costs.

Another saving would be that the farmers could raise their cattle on cheap grass (which usually gives tough, lean meat) instead of expensive grain (which makes the meat more tender). This would be good for the environment, as land could be released from growing grain for cattle.

Another advantage would be for the military.

In the United States, the law forces the military to buy the least expensive cuts of meat, which are also the toughest. The Army would like to use the shock wave treatment, to make tough meat tender. This is no real surprise, seeing as for a very long time already the military have been using explosives to tenderise meat. Unfortunately, the meat they aim their explosives at starts off as a walking, talking human, and ends up as a pile of broken flesh.

TENDERISE MEAT

There are three main ways to make meat more tender: make the *sarcomeres longer*; *rupture the myofibrils*; and *rupture the connective tissue*.

There are two main ways to *make the sarcomeres longer*: stop them from undergoing "Cold Shortening", or stretch them.

"Cold Shortening" is the process by which the cold makes the sarcomeres contract and become tougher. If the animal has lots of subcutaneous fat (fat under the skin), this acts as an insulator and slows the temperature drop, reducing the tendency of the sarcomeres to "cold shorten". Another way to prevent cold shortening is to keep the carcass at 16°C (61°F) for 16 hours, immediately after the animal has been slaughtered. Yet another technique that stops the sarcomeres from shortening is zapping them with electricity — 550 Volts and 2–6 Amps, pulsed 15 times in one minute. This rapidly makes the muscle very acid, and rigor mortis starts much sooner. (In humans, rigor mortis, where the body becomes as stiff as a wooden board, begins about three hours after death, and is fully developed after 12 hours.) Once rigor mortis has set in, the sarcomeres won't shorten.

The other way to lengthen the sarcomeres is simply to stretch them. One method is the "Texas A & M Tenderstretch" which suspends the animal by the obturator foramen (a hole in the pelvis). Another is to use Stouffer Stretching Devices. These are variations of the Texas A & M Tenderstretch, with added clamps and stretching rods.

Another way to increase the tenderness is to disrupt or *rupture the myofibrils*. There are three main ways to do this.

First, you can do this by increasing the activity of internal enzymes that already exist in the meat (calpains or cathepsins). There are several ways to do this. You can age the meat by hanging it for 1–6 weeks in the low temperature range of 0–3°C (32–37°F). It can also be hung, after rigor mortis, at 20°C (68°F) for 24 hours. This is as effective as storing it at 2°C (36°F) for 14 days. You can also increase the activity of the enzymes by hanging the meat, before rigor mortis, at 16°C (61°F) for 16 hours. Electrical stimulation makes the meat more acid, and directly causes the release of cathepsins. Finally, infusing calcium chloride into the muscles before rigor mortis will make the calpains more active.

Second, you can add enzymes, such as those from tropical plants, to disrupt the myofibrils. These include Swift's Pro-ten (injected into the live animal), and Adolph's Meat Tenderiser (sprinkled on the meat).

Commonly used enzymes from tropical plants include papain (papaya), bromelin (pineapple), and ficin (fig).

The third way to disrupt the myofibrils is by mechanically tearing them apart. For example, powerful electrical stimulation of the muscle will trigger violent contractions in the muscle, which rip the myofibrils apart. Myofibrils can also be ruptured with direct mechanical damage from blades or needles, such as chopping, cubing, dicing, grinding or scoring.

The third way to make the meat more tender is to *tear apart the connective tissue.*

Once again you can use acid (salt and vinegar marinade) or external enzymes (rhozyme from the fungus *Aspergillus oryzae*, or enzymes from tropical plants). Another way is to sever the connective tissue proteins using blades or needles. You can convert the collagen in the connective tissue to gelatine by cooking the meat for a long time with steam.

COOK ME TENDER

Meat has two major sets of proteins: collagen and the contracting proteins actin and myosin.

The major protein in connective tissue is collagen. Temperatures above 71°C (160°F) make the collagen proteins break down into gelatins. A meat rich in connective tissue is tough. But if you keep such a meat moist while you heat it above 71°C (160°F), it will become tender.

The proteins in the myofibrils are actin and myosin. They get shorter, and so make the meat tougher, at temperatures around 66–77°C (151–171°F). But above 77°C, the actin and myosin are completely broken down, and so the tenderness begins to improve.

The ideal temperature at which to cook your meat depends on how much connective tissue it has.

If your meat has lots of connective tissue, you should cook it slowly at temperatures above 77°C (171°F), so that the actin and myosin in the myofibrils and the connective tissue all break down.

But if your meat doesn't have much connective tissue, you should cook it at under 66°C (151°F), so that the actin and myosin don't get short and make the meat tougher. It doesn't really matter if the connective tissue doesn't break down and gelatinise, if there isn't very much of it.

MEAT CHANGES COLOUR

Why does meat change colour when you heat it up?

It changes colour because the molecule that carries oxygen breaks down. The colour of this molecule depends on its shape.

In blood, the molecule that carries oxygen is called haemoglobin ("haem" means "iron", and "globin" refers to the "globular" shape of the molecule). When it's loaded with oxygen, the molecule gives blood its characteristic bright cherry red colour. But as it loses oxygen, it changes shape, and gives de-oxygenated blood its dark red colour.

In muscle, the molecule that carries oxygen is a very similar molecule called myoglobin ("myo" means "muscle"). This molecule gives meat loaded with myoglobin a characteristic red colour. (Meat that does not have lots of myoglobin, such as chicken meat, is much lighter in colour.) As meat is heated, the myoglobin molecule breaks down, and so the meat changes colour from the pink around 55°C (131°F) of quite rare to the grey-brown colour 80°C (176°F) of well-done.

David Bortin, "Garbage-Can Superiority", *Science News*, Vol. 154, 8 August 1998, p 83.

J. Mark Loizeaux and Douglas K. Loizeaux, "Demolition by Implosion", *Scientific American*, October 1995, pp 120–127.

Steve Mirsky, "Tender Is the Bite", *Scientific American*, January 1998, p 22.

Janet Raloff, "Ka-Boom! — A Shockingly Unconventional Meat Tenderizer", *Science News*, Vol. 153, 6 June 1998, pp 366–367.

M.B. Solomon et al., "A Research Note — Tenderising Callipyde Lamb with the Hydrodyne Process and Electrical Stimulation", *Journal of Muscle Foods*, Vol. 9, 1998, pp 305–311.

M.B. Solomon, J.B. Long and J.S. Eastridge, "The Hydrodyne: A New Process to Improve Beef Tenderness", *Journal of Animal Science*, Vol 75, 1997, pp 1534–1537.

Morse Solomon, "The Hydrodyne Process for Tenderizing Meat", *Reciprocal Meat Conference Proceedings*, Vol. 51, 1998, pp 171–176.

H. Zuckerman and M.B. Solomon, "A Research Note — Ultrastructural Changes in Bovine Longissimus Muscle Caused by the Hydrodyne Process", *Journal of Muscle Foods*, Vol. 9, 1998, pp 419–426.

REFERENCES

GLASS – FLOWING LIQUID STRONGER THAN STEEL?

There's a big myth about glass, accepted by people as far apart as North and South America, Europe and Australia. Even some university science textbooks claim that "*Glass is a liquid that flows very slowly. To prove it, just look at old windows — they're thicker at the bottom, because the glass has flowed downhill.*" But this isn't true; even science has its myths!

Glass is amazing stuff and does deserve to be the stuff of myths. You can see through it, and it's stronger than steel (but air makes it weak).

Thanks to lenses made of glass, we can see tiny bacteria and distant stars. In modern telescopes, we have glass mirrors as big as a house — 8 metres (26 ft) across. Spray the glass with shiny aluminium, and you have a mirror that can see almost to the edge of the universe.

How to Make Glass

Glass is just transparent rock.

Rock weathers into sand. The main chemical in sand is silicon dioxide (silica, SiO_2). Melt sand, and you get glass you can see through.

There are two disadvantages to making glass from sand alone. First, the melting point of sand is over 1700°C, and you need a lot of expensive energy to generate a temperature that high. Second, the molten sand is very viscous (like very thick honey). Air bubbles can't escape from between the individual grains of sand, and they get trapped inside the glass. So the glass is less transparent.

The solution is to add more oxygen atoms to the sand. These atoms weaken the chemical bonds between the oxygen and the silicon. Then the ingredients will melt at a lower temperature, and the molten liquid is less viscous, so the air bubbles can escape.

For example, if you add 25% sodium carbonate (Na_2CO_3) to 75% sand, these extra oxygen atoms drop the melting point of the mix down to a more reasonable 850°C. (Heating the sodium carbonate, Na_2CO_3, drives off the carbon dioxide, CO_2, leaving behind sodium oxide or Na_2O.)

Unfortunately, glass made from just sand and sodium carbonate will dissolve in water — in fact, it's called "water glass".

Early glass technologists found that adding limestone or lime (which contains calcium oxide or CaO) makes glass insoluble in water. Another benefit of adding lime is that it adds more oxygen atoms, which further weaken the chemical bonds between the oxygen and silicon atoms.

Glass — Not a Crystal

In general, there are two types of solids — crystalline and amorphous. ("A-" means "not", and "morphous" means "shape", so amorphous solids have no regular shape or pattern at an atomic level.)

In a crystal, the atoms are locked into symmetrical patterns that repeat themselves and the bonds are all the same. Crystals melt suddenly over a narrow temperature range.

Amorphous solids have atoms arranged much more loosely. Some atoms are close to each other, while others are further apart. The chemical bonds between neighbouring atoms all have different strengths. The weak ones will break at a low temperature, while the stronger ones will break at a higher temperature. So amorphous solids soften gradually over a wide temperature range.

Lead crystal glass is not a crystal and has no crystalline properties, despite the name. It was invented in 1674 by George Ravenscroft in England. He was trying to make a clear glass that was not greenish like the other English glasses of the time. He added red lead oxide to his ingredients and made a glass that was as clear as natural rock crystal, so he called it lead crystal glass.

So glasses are amorphous solids. They are stiff frozen liquids that have forgotten how to flow.

How Did We Invent Glass?

An ordinary campfire is usually no hotter than 700°C (1300°F). So lighting a big fire at night on a sandy beach probably wouldn't leave little lumps of glass in the morning.

We really don't know how mankind invented glass, but there are two popular theories.

The first one says that someone accidentally did it by heating ores to make metals. Some of the molten by-products have a glassy finish, and this supposedly inspired people to try to make a glass. The second theory says that people were already using ovens to make ceramics, so they just modified the process to make glass.

History of Glass

Nature has been making glass for many millions of years. Examples of this natural glass include volcanic glass, and glass found in some plants, like on the tip of the stinging nettle.

First Flowering — Egypt

Glass beads from Egyptian tombs some 4500 years ago are among the earliest 100% glass objects that we have.

But the earliest glass vessels come from the 18th Dynasty of Egypt (1580–1358 BC). These vessels were often moulded around a central core of sand or clay, which was removed after cooling. The molten glass was sometimes rolled onto the central core, or wrapped around it as coloured threads.

The ancient Egyptians got their silica from sand or crushed pebbles. They got soda ash from "natron", which came from the vast dried-out lakes of Wadi el Natrun between Alexandria and Cairo. (However, the Mesopotamians got their soda ash from the ashes of the "naga" plant.) Lime came from limestone, chalk or burnt sea shells.

Rome to the Islamic World to Europe

Glassmaking developed into a major industry in Roman times — helped by the invention of glass-blowing. By 400 BC, people had glass windows, glass storage containers and glass urns to rest their funeral ashes in. When they went to the Colosseum, they could even buy special souvenir glass cups, embossed with the names of their favourite gladiators. (This is the earliest Great Moment in Merchandising that we know of.)

The Roman Empire fell, and, during the Dark Ages, the art and science of glassmaking shifted to the Islamic world, where it flourished. After the turn of the millennium, it shifted back to Europe, to the artisans of Venice and the makers of stained glass.

They did beautiful things with glass, but they didn't realise that it was stronger than steel.

Glass — Stronger than Steel

Glass is one of the strongest materials known to the human race. But there are a few conditions you need to set up to keep it super-strong. First, begin with glass that has no flaws. Second, keep the glass in a high vacuum to stop air molecules from smashing into it and water vapour from attacking it. Then try to pull the glass apart. You'll find that this glass is about 10 times stronger than most commercial metal alloys and 50 times stronger than mild steel.

Glass — Weak and Brittle

So why is common glass so weak and brittle?

The atmosphere carries abrasive particles, corrosive chemicals and water vapour to the glass. Together, they cause tiny cracks in the surface. These cracks weaken the surface of the glass and eventually spread deeper.

You might have seen a crack slowly spread across the windscreen in your car over a period of years. Cracks can move in

Man of steel
v
Woman of glass

glass at many different speeds — as slow as a trillionth of a centimetre per hour, or as fast as hundreds of metres in a second. Water can speed up the growth of cracks in glass by more than a million times, because it weakens the structure of the glass at the very tip or root of the crack.

So if glass is so strong in the laboratory, but weak in real life, how can engineers hope to build bridges out of it? The answer lies in very skinny, but very strong, fibres of glass.

Glass Fibres + Matrix = Fibreglass

Back in the 17th century, Robert Hooke suggested that glass-blowers might copy how the silkworm makes silk. He proposed that they force molten glass through very small holes to make very thin fibres of glass.

In 1713, the French scientist René de Réaumur showed a crude fibreglass fabric to the Academy of Science in Paris. In 1836, another Frenchman, Dubus-Bonnel, was awarded a patent for his process of spinning and weaving glass fibres. In 1893, the American Edward Drummond Libbey managed to spin fibres of glass, and combined them with silk. Unfortunately, the resulting fabric was so stiff that it broke easily. Fibreglass was reinvented in 1931 by the Owens Illinois Glass Company. At first, it was used only for heat insulation.

An early use of fibreglass for its strength was in the body panels of the 1953 Chevrolet Corvette. Soon fibreglass was used in boats as well. These skinny fibres can be extraordinarily strong — 66 666 tonnes per square metre (1 000 000 psi). However, they have an enormous surface area relative to their mass. The fibres are usually sprayed with protective chemicals, or buried in a matrix of plastic or epoxy to protect them from the air so they can retain their strength.

Today, fibres of glass are used for flexible sealing tape, fireproof curtains, heat insulation, body panels in boats and cars, filters, and even to make roads and bridges.

Glass Roads and Bridges

Frieder Seible is a structural engineer and director of the Charles Lee Powell Structural Research Laboratories at the University of California in San Diego. He has been designing and building advanced composite structures for a long time. He has successfully replaced asphalt with black glass fibres in the surface of a road used by heavy construction trucks — and it's surviving very nicely.

WINDOW = WIND'S EYE

In ancient Scandinavia, houses didn't have glass windows. Instead, they had a hole, or "eye", in the roof. The wind would come through this "eye", so they called this hole the *vindr auga* — meaning the "eye" for the "wind", or the "wind's eye".

But today, a glass "wind's eye" will block the wind.

PLATE GLASS

It's actually quite difficult to make smooth, flat plate glass. It's amazing that even though we humans have been making glass for 4500 years, we learnt to make cheap flat glass only in 1959.

The Romans made a crude plate glass by heating the glass and pulling it into a sheet. There probably wasn't a great need for smooth, flat window glass, because of the mild Mediterranean climate.

Around 600 AD, glassmaking experts set up their workshops along the Rhine River. They were given the title of "gaffer", from "learned grandfather". It was a term of respect, reflecting their hard-won knowledge and experience.

These gaffers had a few techniques of making window glass. Each method produced a wavy plate glass with a rough finish.

In the "cylinder" method, the gaffer would blow molten glass into the shape of a cylinder, which would then be cut open to make a glass plate.

The "crown" method used centrifugal force to turn a spinning fat blob of hot glass into a spinning thin disc. A 4-kg (9-lb) blob of hot glass (called a "gob") was kept warm in a "flashing furnace". It was attached to the end of a long rod. The gaffer spun the end of the rod while he stood several metres away from the heat of the furnace. As the hot gob spun at high speed, it would suddenly open with a sound just like "*quickly expanding a wet umbrella*". The hot gob would then change into a circular disc of plate glass about 150 cm (60 in.). The disc of glass was thicker right in the middle, and also on the circumference. The "splitter" would then cut out the smoothest and flattest sections. Even after it was polished, it was still a fairly wavy sort of glass.

In 1687, Bernard Perrot of Orléans patented a new way to roll plate glass. This was the first high-quality plate glass. He poured the hot liquid glass onto an iron table and then spread it out with a metal roller.

"Mechanical rolling" was introduced in 1918. Molten glass was rolled between rollers in the same way that sheet metal is rolled. Once again, it needed to be polished, and there was still obvious waviness in the glass.

In 1937, the Pilkington method of simultaneously grinding and polishing plate glass on both sides was introduced. But this type of glass was still expensive.

Finally, in 1959, Alistair Pilkington of the Pilkington Glass Company invented the "float glass" method. He had noticed that when he washed the dishes in hot water, a layer of liquid fat would float on the top of the water. It took him seven years and $16.5 million to get the "float glass" method right. Molten glass is poured onto a bath of molten tin at 1050°C (1920°F). The glass spreads to be perfectly flat on both sides, with no need for further polishing. It is removed when it cools down to 650°C (1200°F).

Finally, plate glass that was both inexpensive and high quality became available.

making flat plate glass with the "crown" method

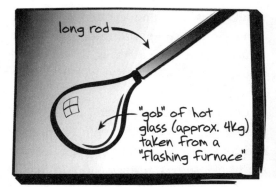

long rod

"gob" of hot glass (approx. 4kg) taken from a "flashing furnace"

1. the molten glass

the "gaffer" enlarges the glass "gob", assisted by a young apprentice

young apprentice   gaffer

2. enlarging the glass flask

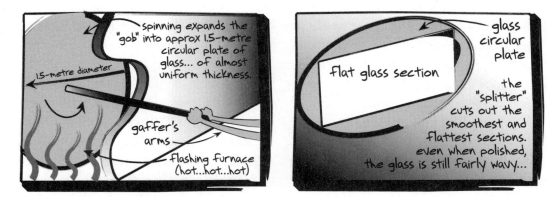

spinning expands the "gob" into approx 1.5-metre circular plate of glass... of almost uniform thickness.

1.5-metre diameter

gaffer's arms

flashing furnace (hot...hot...hot)

3. spin baby spin!

glass circular plate

flat glass section

the "splitter" cuts out the smoothest and flattest sections. even when polished, the glass is still fairly wavy...

4. presto! – flat glass

One of Seible's colleagues, John Kosmatka, has already designed and built a Composite Assault Bridge 4 metres wide and 15 metres long out of glass. This portable bridge is used to carry military vehicles across ditches. The special carrying vehicle is just a tank without a turret. The bridge weighs only 4 tonnes, but can carry a 70-tonne tank! In fact, it's half the weight of the equivalent steel bridge, so the Portable Bridge Carrier can now carry two bridges. The United States Army loves this idea, because it instantly doubles the number of Assault Bridges they can mobilise.

Part of the reason that engineers are trying to make bridges out of glass and other non-metallic materials is that the civil infrastructure of much of the world is in a really bad state. For example, steel rods buried in concrete corrode and swell as moisture and salts ooze into the structure. In the United States alone, there are 600 000 bridges, but 42% of them are obsolete, or, at the very least, in need of major repairs. It will cost about $US50 billion to fix them all. Throughout the world, bridges are corroding faster than we can repair them.

Maybe it's time for a new approach.

Smart Glass Roads and Bridges

Composite bridges can be much lighter and stronger than those made of steel and/or concrete. Composite materials also have another advantage — they can repair themselves. Engineers propose to install small tubes of liquid epoxy between the fibres of the composite materials. If the composite moves and cracks, the tubes will break, and liquid epoxy will leak out, repairing the crack.

Composite materials can also monitor themselves. You can install optic fibre cables and constantly shine light through them. If the bridge suffers an overload and bends, the frequency of the light will change, which could set off alarms.

However, we need to build roads and bridges from these new materials to test how they survive in the real world. Maybe we'll find that sunlight weakens these new composites.

A major interstate highway runs right through the middle of the campus of the University of California at San Diego. Engineers are now working on a $10 million, 140-metre-long (460-ft) composite material bridge that will carry four lanes of traffic over this highway. It will be five times lighter than a steel and concrete bridge.

This will be the first bridge in the world to be built using new composite materials. It is engineering in a *glass* of its own!

Stained Glass Cathedrals

But getting back to around the 1100s, glass suddenly took on a new importance — in religion rather than engineering.

Builders had worked out how to make bigger Gothic cathedrals, but wherever you have a pointed roof, it tries to push the walls apart. To solve this problem, the builders invented external supports called "flying buttresses", that lean up against the cathedral wall. These flying buttresses now carried part of the load of the heavy roof. This meant the builders could make holes in the walls where they could install stained glass to impress the peasants with religious pictures made from magically coloured light. After all, God's first creative command was, *"Let there be light"*.

It was the stained glass in these cathedrals that most probably led to the story that glass would flow slowly, like a very thick liquid. This made its way into urban myth, science textbooks, and even some popular encyclopaedias.

the evolution of the flying buttress

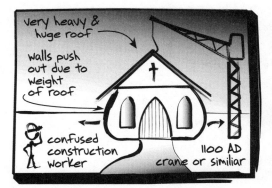

1. a big gothic cathedral being built...in 1100

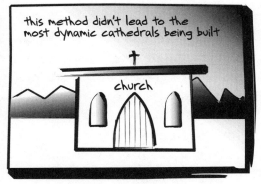

2. one solution...

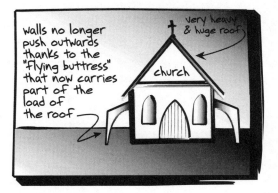

3. the flying buttress is now here...

TEMPERED GLASS — HEAT TREATMENT

Tempering is one way to make glass strong.

To understand tempering, you have to know the difference between compression and tension. Compression is when you try to squash something, while tension is when you try to pull it apart. Brick and concrete are very strong when you compress them, but theycrumble and break easily when you try to pull them apart. A steel wire is very strong when you try to pull its ends apart, but it is weak when you push them together — it buckles.

To make tempered glass, you rapidly cool a sheet of molten glass.

First, the outside will shrink as it cools. But then it can't shrink any further, because of the still molten glass inside, so it stops shrinking and freezes solid. Because the glass on the outside gets frozen while it's trying to squash the glass on the inside, it gets locked into a state of compression. (This compression force is typically 350 atmospheres.)

As the inside molten glass continues to cool, it tries to shrink away from the outside glass while still attached to it. It freezes while trying to pull away, so the inside gets frozen in tension.

This is tempered glass — the outside in compression, the inside in tension.

Glass is about 25 times stronger when it's squashed (compression) than when it's pulled (tension). If it's going to break, it will do so in tension.

Apply a force to the tempered glass. If it's a compression force, the glass doesn't break, because it's very strong in compression. If it's a tension force, the tension force is reduced by the amount of the compression force already locked into the glass, so the glass won't break. (Of course, a very great tension force will overwhelm the compression force in the glass and break it.)

And why doesn't the glass on the inside, which is in tension, break? Because it's protected from the atmosphere, so it has no flaws and is very strong.

When tempered glass does eventually break, it shatters into thousands of tiny harmless fragments, rather than long, sharp, dangerous shards.

tempered glass - the strong one!

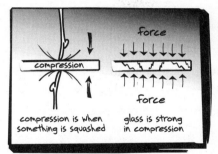

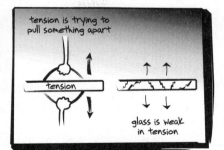

1. stuff to know first...

compression is when something is squashed

glass is strong in compression

2. tempering the glass...

tension is trying to pull something apart

glass is weak in tension

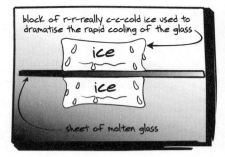

block of r-r-really c-c-cold ice used to dramatise the rapid cooling of the glass

ice

ice

sheet of molten glass

3. tempered glass recipe

the outside shrinks as it cools... but then can't shrink any further, because of the molten glass inside... outside glass freezes in compression.

molten glass inside

cooled outer layer is still in compression

4. the inside story...

as the soft centre of molten glass cools further, it tries to shrink away from the solid outside glass... & freezes while doing this...it then remains in tension!

→ tension (stretched) inner ←

compressed (squashed) outer layer

5. more inside gossip!

glass on the inside is protected from the atmosphere...it has no flaws and is very strong.

→ tension (stretched) inner ←

compressed (squashed) outer layer

6. all the secrets revealed

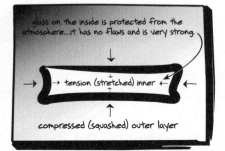

top layer is in compression & is able to accommodate added compression forces

heavy anvil (force applied to tempered glass)

the glass doesn't break as it's very strong in compression.

7. tempered glass in action

bottom layer starts off in compression... the glass is curved & this tries to pull the glass molecules apart (= tension)...

heavy anvil (force applied to tempered glass)

the glass doesn't break as it's still in compression.

COMPRESSION - LITTLE TENSION = SOME COMPRESSION

8. the temperamental end!

TEMPERED GLASS — CHEMICAL TREATMENT

You can also temper glass with chemicals. Cool it rapidly by dipping it in a solution of potassium chloride (KCl). The potassium atoms boot out and replace some of the sodium atoms near the surface of the glass. Potassium atoms are bigger than sodium atoms, so they put a compression force into the outer layer of the glass.

SAFETY GLASS

Today, we make safety glass by joining two sheets of thin glass (2–3 mm or 0.08–0.12 in) with a transparent membrane (usually polyvinyl butyral) using pressure and heat.

Safety glass was discovered accidentally by a French chemist, Édouard Benedictus, in 1903. An assistant had put a liquid plastic called cellulose nitrate into a flask. It evaporated away, leaving behind a clear invisible coating. Thinking that it was already clean, he put it away without cleaning it. Benedictus accidentally knocked this particular flask to the floor. It broke, but the pieces stayed together.

By coincidence, a Paris newspaper ran a major article that week about automobile accidents, and how many of the victims had been seriously injured by long shards of sharp windscreen glass.

Benedictus wrote in his diary: "*Suddenly there appeared before my eyes an image of the broken flask. I leapt up, dashed to my laboratory, and concentrated on the practical possibilities of my idea. By the following evening, I had produced my first piece of Triplex [safety glass] — full of promise for the future.*"

The car manufacturers didn't want to spend more than they were forced to, so they refused to use his Triplex Safety Glass. It was first widely used in World War I in the lenses of gas masks. But over the years, the car manufacturers gradually adopted his safety glass.

"Proof" 1 — Old Glass Is Thicker at the Bottom

That is true — most of the time. When you look at old stained glass around the world, and even glass from windows in old houses, you'll find that most of it is thicker at the bottom. But it's not because the glass flowed downhill. It's because we have learnt to make flat plate glass only very recently. When the older uneven glass was put into a window, the window-makers usually put the thick section at the bottom. After all, when

DIFFERENT GLASSES

The most common glass today is soda–lime–silica glass — made from 15% soda ash (sodium oxide, Na_2O), 10% calcium oxide (CaO), 70% silica sand (silicon dioxide, SiO_2) and 5% various other oxides. This is a good all-round glass that is easy to soften and shape, yet resists chemicals.

But there are an infinite number of possible recipes for making glass. Between 1850 and 1960, the Corning Glass Works in the United States had tested 65 000 recipes, and was still testing new recipes at the rate of 30 per day. To make every type of existing glass, you need to use half of the elements in the periodic table.

Lead glass is made from silica sand and lead oxide. It has a high electrical resistance and a high refractive index. It is used in tableware, art, and electrical applications.

Borosilicate glass contains about 10–15% boron oxide (B_2O_3), along with silica sand and soda. This glass is often called Pyrex. When you heat it, it expands only one third as much as common soda–lime glass, so it's resistant to thermal shock. Pyrex is also very resistant to attack by chemicals, so it's used in cookware and in chemical laboratories. In its early days, it was used to deal with the problem of "thermal fracture", when cold rain hit the hot glass in the front of a railway signal lamp.

Solder glass has oxides of lead, boron and zinc. It softens at a low temperature, so it can be used to join different glasses, metals or ceramics.

You can make a glass that hardly expands at all when you heat it — about 100 times less than ordinary window glass. It's made from 74% silicon dioxide, 16% aluminium oxide, 6% titanium dioxide, and 4% lithium oxide.

you construct a building, you usually put the heaviest and biggest bits at the bottom.

Peter Gibson, who has dedicated his life to dismantling and restoring medieval glass windows, agrees. He spoke about this at the International Conference on Industry–Education Initiatives in August 1995 in York, UK. He says that while most of the glass that he has seen was thicker at the bottom, he has seen hundreds of pieces of old plate glass that were thicker at the top.

For example, consider the windows of St Peter's Church in Cheshire, Connecticut. They were installed in 1840. Most of them are thicker at the bottom.

But one sixth are thicker at the sides, while a few are thicker at the top. Obviously the glass did not flow sideways, or uphill — and neither did it flow downhill.

"Proof" 2 — Instructions on Boxes

Evidence for the "fact" that "glass flows slowly" are the instructions printed on the end of the boxes of long glass tubes sold by the Corning Glass Company. The instructions say, "Lay Flat, Do Not Store on End".

Indeed, when you remove a long glass tube, you can sometimes see a slight bow in it — "proof" that is has flowed while it was stored leaning against a wall.

But this bow is an unavoidable part of the manufacturing process. Corning allows a bow of 3.8 mm (0.15 in.) in a 122 cm (4 ft) length. According to the Corning Company, the real reason not to store the box leaning against a wall is to prevent damage to the ends of the glass tubes.

Telescope Lenses Don't Flow ...

Astronomers have been making and using telescopes with large glass lenses for well over a century. They would definitely have noticed if the glass in their lens flowed. Lenses need a shape that is accurate to about one tenth of a wavelength of visible light — roughly 50 billionths of a metre. But not one astronomer has ever complained that their old lens is unusable because the glass has flowed into a new shape.

Astronomers Make Mirrors from Glass ...

Astronomers don't make glass lenses much bigger than a metre or so. When the lens gets too big, the glass in the middle of the lens sags. This is not the glass flowing — it's just sagging. Steel will sag. When you flip the lens upside-down, it will sag the other way — again, just like steel.

Astronomers call their telescopes "light buckets". They go out at night with their light buckets and try to catch as many photons of light as possible. A bigger light bucket means that you catch your photons more rapidly, so you can quickly point your light bucket somewhere else to take another picture. But big lenses sag, so astronomers shifted to mirrors for really big light buckets.

The advantage of a mirror is that you can make the backing of the mirror as big and stiff as you want. Glass telescope mirrors don't use the *transparency* of glass. Instead, they use its strength, rigidity, lack of change with temperature, and ability to be ground and polished into the exact

RECYCLING GLASS

In the United States, glass makes up 6% of all solid wastes. About 35% of glass is recycled (most is buried in landfill). It takes 25% less energy to liquefy recycled glass than to liquefy virgin ingredients.

In Holland and Switzerland, the recycling rate is 50%, but in Britain it's only 12%.

SUNBURN THROUGH GLASS

In general, you won't get sunburnt through glass. Glass (but not most plastic) stops a lot of the UV light less than 350 nm. If your skin does go red after sunlight that has passed through glass has fallen on it, it's most likely a heat effect, not a sunburn.

But it is possible to get sunburnt from even the tiny amount of UV that comes through glass, if you are of Irish/Scottish descent, with red hair, freckled skin, and light-coloured eyes.

shape you want. Several big telescope mirrors have been in use for over half a century. The professional astronomers who operate them definitely would have noticed if the glass had flowed.

The Answer

Practically all plate glass made before 1959 had some degree of waviness in it. It was uneven in thickness. As mentioned before, the glaziers would install the glass with the thick bits at the bottom and the thin bits at the top to enhance the building's stability. So old windows usually are thicker at the bottom, but that doesn't prove that glass flows.

Nobody has ever been able to measure the flow of glass. But this hasn't stopped scientists from trying to work it out theoretically.

According to the mathematical calculations of Edgar Dutra Zanotto, from the Department of Materials Engineering at the Federal University of São Carlos in Brazil, glass can flow — but really incredibly, incredibly slowly.

In May 1998, he wrote an article in the *American Journal of Physics*, where he worked out how fast glass can flow. He looked at how fast molten glass flowed, and then extrapolated from the curves on a plotted graph how fast it would flow at lower temperatures. According to his calculations, if you heated the glass to 414°C (777°F), it would flow a visible amount in only 800 years. But if you wanted to see glass flow at room temperature, you would have to wait at least 10 000 million million million times the age of the universe!

However, Prabhat K. Gupta from the Department of Materials, Science and Engineering at Ohio State University pointed out that Dr Zanotto had calculated only the maximum time for the glass to flow. This was because he had used the concept of "equilibrium viscosity" (I won't go into it here). So Zanotto and Gupta wrote a combined paper, again in the *American Journal of Physics*, using "isostructural viscosity". This time, they got a smaller time — only 100 million million times the age of the universe.

Dr Yvonne Stokes from the University of Adelaide took a different mathematical approach and used the equations of fluid dynamics. Her calculations show that any perceptible thickening of the glass would take at least 10 million years. However, if she had used Zanotto's values for "isostructural viscosity", she would have ended up with times closer to his.

In fact, Stokes' mathematical model had a surprise outcome. The flowing glass would give you a pane of glass that was the original thickness at the top, but thicker at the bottom. This extra thickness came from a reduction in height. The glass pane would pull away from the top, leaving a gap. A 100-cm (39-in.) high sheet of glass would (given sufficient millions of years) shrink down to 98 cm. Really old glass would not be thin at the top — it would have holes in it!

Using either mathematical approach, you are looking at millions of years for glass to flow at room temperature.

So waiting for glass to flow at room temperature is enough to make your eyes glaze over — and a real pain as well.

REFERENCES

Robert H. Brill, "Ancient Glass", *Scientific American*, November 1963, pp 120–130.

Charles H. Greene, "Glass", *Scientific American*, January 1961, pp 92–105.

S. Muspratt, *Chemistry Theoretical, Practical & Analytical as Applied and Relating to the Arts and Manufacture*, Vol II, Mackenzie, 1860, pp 21–216.

Robert C. Plumb, "Antique Windowpanes and the Flow of Supercooled Liquids", *Journal of Chemical Education*, Vol. 66, No. 12, December 1989, pp 994–996.

Jim Wilson, "Glass Bridges", *Popular Mechanics*, December 1997, pp 74–77.

Edgar Dutra Zanotto, "Do Cathedral Glasses Flow?", *American Journal of Physics*, Vol. 66, No. 5, May 1998, pp 392–395.

GOBBLING WHIRLY BIRDS

Scientists have found animals that can swim, squirm, walk, jump, hop, flop, flap, flip and even fly. But in 1996, they found the only animal that spins — a tiny seabird. Because this bird pecks as it spins, and because it spins so fast, it's the fastest feeder on our planet — 180 mouthfuls every minute!

This study was carried out by Bryan Obst, William Hammer (from the Department of Biology at the University of California in Los Angeles) and their colleagues. This spinning animal is a wading seabird or shorebird, called the phalarope. The phalarope is a little critter — about 15–25 cm (6–10 in.) long, with a slim neck. To help them swim better, they have toes with lobes (to increase the surface area of their feet).

The Red-Necked Phalarope

There are three species of phalarope, but the one that the scientists looked at is the red-necked variety, also called the northern phalarope (*Phalaropus lobatus*). It breeds up around the Arctic Circle and then, come winter, nicks off to the tropics. In these warmer waters, they're called the "sea snipe" (but the ornithologists, or bird scientists, still call it a phalarope). In summer phalaropes have red and soft grey marks, but in winter these change to grey and white.

They're very unusual birds, with skills including sex role reversal, super fast feeding, and spinning.

Phalaropes Run a Non-Traditional Household

The female is brightly coloured, larger than the male, and is aggressive in courtship and in defending her territory. The male does all of the incubation, by sitting on the egg.

Once the baby is born, the female helps out a bit with the feeding, but then all the females leave to migrate southward. It's left to the older males to lead the younger ones down south.

A few other birds, such as most jacanas and the painted snipe, also have sex role reversal, but in general this phenomenon is rare.

Fastest Feeder

What is really amazing about the phalarope is that it feeds faster than any other animal on the planet. It has a routine where it can find its prey, peck at it, grab it with its bill and swallow it — all in a third of a second!

To manage to feed this fast, the phalarope spins like a whirling dervish — this creates a whirlpool in the water that sucks the underwater food to the surface.

The birds spin about once every second, or 60 times per minute. Each 360° spin requires about seven or eight separate kicks from their lobed toes. The toes spread out for the push stroke, and then fold up for the return stroke. This spin generates a whirlpool, or vortex, in the water, which sucks the food up to the surface. The phalarope leans inward, keeping its bill over the centre of the whirlpool, where it gobbles up the food.

When they spin, phalaropes kick so vigorously with their lobed toes that they actually make a depression, or hollow, in the surface of the water. Deeper water wells up to fill the gap. The physicists call this "an upward momentum jet". Once this deeper water hits the surface, it spreads outwards and parallel to the surface, in a layer about 2 cm (0.8 in.) thick.

In one experiment, the scientists placed a phalarope in water that had been stained with a dye, directly over a dead shrimp in a dish. As the bird began to spin and feed, the scientists could see that "a green tornado-like tube of dye and prey rose upward from the dish, rotating opposite to the rotation of the bird as it began to feed". The beauty of spinning is that it concentrates the food from a wide area into a small feeding zone. Spinning works really well when the movements of the prey are slowed down by the cold and when the prey is in layers in shallow water (although phalaropes will spin in water that is many fathoms deep).

HOW TO SPIN A PHALAROPE

The scientists got the red-necked phalarope to spin by placing some brine shrimp in an open dish under the water. The phalarope could easily reach down to swallow the brine shrimp by dipping its bill and head completely underwater.

Then the scientists gradually increased the water level, until the phalarope began to spin. The scientists had already added a harmless yellow dye to the water to show the movement of the spinning whirlpool. When they shone an ultraviolet light on the yellow dye, it glowed a brilliant green.

Phalaropes Don't Get Dizzy

You've probably seen a ballerina perform a pirouette. They may spin their body continuously, but they keep their eyes fixed on a single point in the audience and snap their head around once per spin. This way they don't get dizzy (and it looks cool).

The feeding phalarope performs an equally cool trick. The body spins continuously, but the head snaps around in separate 45° bursts. And 180 times each minute, they will find, peck at, and swallow their prey. They feed faster than any other animal known to mankind. They really stretch the concept of "fast food" to its limits.

Phalaropes also have another clever trick. Each spin is slightly offset from the spin before it, so the birds continually trawl over new areas.

It takes a lot of energy to spin — about four times the phalarope's basal metabolic rate at rest. These birds do not spin when there is lots of food close to the surface, or where they can get at food just by dipping their bill and head under the water. Nor do they spin when it's windy, because the waves will interfere with their upwelling whirlpool. Only a smallish bird can spin fast enough to generate a decent whirlpool that will suck up its dinner, so we won't find large spinning whirly birds.

They say that the early bird catches the worm, but, with the phalarope, it's the whirly bird that whips up a storm.

REFERENCE

Bryan S. Obst, William M. Hammer, et al, "Kinematics of Phalarope Spinning", *Nature*, Vol. 384, 14 November 1996, p 121.

THERAPEUTIC TOUCH

People get sick — and so we invented medicine. But medicine changes with time.

One very new branch of medicine is called Therapeutic Touch (TT). "Therapeutic" means "making better". Practitioners of TT claim that they can "feel" an imbalance in your "Human Energy Field" (HEF) by running their hands over your body. They further claim that they can use their hands to fix your HEF and bring you back to good health.

These are extraordinary claims. Extraordinary claims need extraordinary proof. But there are no such proofs.

In fact, the opposite is true. A simple and elegant experiment, carried out by a nine-year-old girl, suggests that the claims of Therapeutic Touch are out of touch.

Therapeutic Touch — A Snapshot

Over 100 000 people across the world have been trained in Therapeutic Touch. Over 40 000 of them are in America. It's claimed that TT is taught in over 100 colleges and universities in 75 countries.

It's also claimed that TT is used by nurses in at least 80 hospitals in North America. In Toronto, Ontario, you can tap into the local network of some 600 practitioners by dialling 65-TOUCH on your phone. A hospital in Denver (Presbyterian St Luke's Hospital) has a "Department of Energy" to support its own practitioners of TT. Bristol Hospital in Connecticut runs its own 15-hour course in TT. About one quarter of the health care staff have taken this course.

Origin of Therapeutic Touch

Therapeutic Touch was founded in the early 1970s by a group of people headed by Dolores Krieger.

Krieger was a registered nurse with a Ph.D., and was employed as a faculty member at the Division of Nursing at New York University. She claimed that people radiated a strange "Field" or "Life Force". She claimed that this "Life Force" could literally be felt by other people. If you're unwell, this "Life Force" supposedly changed. The TT practitioner is supposed to be able to "feel" the twisted or abnormal "Life Force" coming from the sick person, and then somehow change this "Life Force" with their bare hands, and so make the sick person better.

None of the sciences (life, physical, engineering, military, etc.) have ever measured or proved the existence of this "Life Force". This did not bother Dolores Krieger at all.

Originally, this "Life Force" was called "*Prana*". Today it's called a "Human Energy Field", or HEF.

Dolores Krieger also wrote that the healing effect depended entirely upon the practitioner, and did not need any acceptance or belief by the sick person for it to function. She wrote: "*faith on the part of the subject does not make a significant difference in the healing effect*".

In the early days of Therapeutic Touch, the technique did actually involve touching the patient, whether directly on their skin or through clothes. But as it's practised today, TT involves no physical flesh contact at all — you just feel, and then heal.

Therapeutic Touch, as it is taught today, consists of a few basic steps.

Step 1 — Centering

The first step is "centering". Here the practitioner has to focus their intent (or desire) to make the patient better. This step is a little similar to meditation, and supposedly benefits the practitioner as well.

When practitioners are properly centred, they experience a "*decisive shift in consciousness*". This lets them detect the "Life Energy". Practitioners must remain centred

throughout the entire process of TT, to gain understanding of their own and others' "Life Energy".

Step 2 — Assessment

The second step is "assessment". Here the practitioner works out what's wrong with the patient. The practitioner runs their hands, at a distance of 5–10 cm (2–4 in.), over the patient's body from their head to their feet. They do not touch the patient at all. They attune themselves to the condition of the patient, by feeling what they call "*changes in sensory cues*" in their hands.

During assessment, the practitioner should pick up "*vague hunches, passing impressions, flights of fancy, or, in precious moments, true insights or intuitions*". A "good" assessment will tell the practitioner what to do during the next phases of treatment.

Importance of Assessment

Both the theory and the technique of Therapeutic Touch demand that a practitioner *has* to be able to feel the HEF of the sick person before they can heal them. In the literature, practitioners use words such as "tingling", "throbbing", "hot", "cold", "spongy" and "tactile as toffee" to describe what they feel through their hands when they're "assessing" their patients.

So, apparently these Therapeutic Touch practitioners *really* can feel these "Human Energy Fields", even though these mysterious fields have never been picked up by any machine built by the human race.

THERAPEUTIC TOUCH

Therapeutic Touch was created in the early 1970s by Dolores Krieger, R.N., Ph.D., and Dora Kunz. Dora Kunz claimed she was a *"fifth generation sensitive, and had been clairvoyant since birth"*. Originally, they claimed that the Hindu "vital force" known as *prana* was the source of all TT healing.

Krieger's first study had a very uneven distribution of patients — nine in the control group and 19 in the therapy group. The "science" in her first study was terrible. There was no attempt to compensate for factors that might interfere with the results, such as use of drugs, changes in diet, blood transfusions, time of the menstrual cycle, etc. The studies that followed weren't much better. Soon, all their "scientific" studies claiming to show the good effects of TT had been roundly condemned. So they dumped the *prana* explanation and started groping for a new "scientific" basis for TT.

Luckily for them, around this time the Dean of Nursing at New York University, Martha Rogers, was developing her "Science of Unitary Human Beings". This included the concepts of *"Homo Spacialis"* (humans in outer space being the next evolutionary phase of humans), transcendence, pandimensionality, and a whole lot of other big words, and unprovable pseudoscientific mumbo-jumbo. According to her theory, humans didn't *have* "Energy Fields" radiating out of them — they *were* "Energy Fields".

Rogers later expanded her theory to talk about the *"non-linearity of time"*. She also said that her theory explained both clairvoyance and telepathy, and how you did not need to have any physical contact at all to transfer the energies.

In 1982, Janet Quinn agreed. She wrote her thesis claiming that, in TT, you did not need to touch the patient. As a result, TT turned into TNT (Therapeutic Non-Touch).

Dolores Krieger happily seized on this concept of "Human Energy Fields" to give a "scientific" basis to TT. The "scientific" basis also includes other meaningless collections of random adjectives and nouns such as *"The theoretical foundation of TT is based on the Rogerian Conceptual Model of Integrality which views the person and environment as complementary multi-dimensional energy fields engaged in mutual and simultaneous interactions"*.

In November 1994, *Time* magazine wrote about TT. It said, "*as proof of TT's efficacy, they cite 'scientific' reports in such obscure journals as* Subtle Energies *and* Psychoenergetic Systems, *as well as stories in popular magazines*".

Step 3 — Intervention/ Unruffling

The third step is "intervention". This is where the practitioner begins to "heal" the patient. Here the hands of the practitioners "'repattern' the 'energy fields' of the patient by removing 'congestion', or 'replenishing depleted areas', or 'smoothing out areas of energy' which do not flow well".

The third phase is also known as "unruffling". The energy can either be redistributed to other areas which have too little energy, or completely removed from the body by sweeping it down the legs and off the feet. Of course, the excess energy has to be shaken off their hands using a motion similar to shaking water off fingers.

In this case, the TT practitioner must "wrist-flick". Ramacharaka, in *Psychic Healing*, wrote: "*When the pass is completed, you swing the fingers sideways, as if you were throwing water from them*". People in this field seem sincerely to believe that this energy is powerful stuff and could have bad effects. Powell, in *The Etheric Double*, wrote: "*The operator must take care to throw off from himself the etheric matter he has withdrawn: otherwise, some of it may remain in his own system, and he may presently find himself suffering from a complaint similar to that of which he cured his patient*".

Sometimes energy is transferred from the TT practitioner to the patient. If a "*coldness*" has been felt, then a "*hotness*" must be transferred to relieve the imbalance.

"Science" of TT

There are over 850 reports about Therapeutic Touch in various books and journals. However, only 80 or so of them actually describe clinical research into Therapeutic Touch. Until nine-year-old Emily Rosa's paper, not one of them even tried to test how well a practitioner of Therapeutic Touch could pick up the HEF.

Yes, I feel much better

Most of the studies of TT that have been carried out by TT practitioners suffer from bad science. They have improper or absent double-blinding measures, terrible statistics, poor controls, poor design and poor methodology. They are fundamentally flawed.

Therese Meehan, R.N., Ph.D., is a TT researcher who is very much in favour of TT. However, she is quite open about the fact that there is no science that shows the benefits of TT. She wrote in a letter to the *American Journal of Nursing*: "*What current research tells us is that … there is no convincing evidence that TT promotes relaxation and decreases anxiety beyond a placebo response, that the effects of TT on pain are unclear and other replication studies are needed before any conclusions can be drawn*".

Wallace Sampson, M.D., editor of the *Scientific Review of Alternative Medicine*, wrote in the Spring 1998 issue: "*A recent review of the literature shows that there is no convincing evidence that the alleged healings by Therapeutic Touch are anything more than the placebo effect*".

$742 000 Experiment

One of the early scientific tests of TT was carried out on 14 November 1996, offering prize money of $US742 000.

The experiment was jointly sponsored and conducted by the James Randi Educational Foundation and PhACT (Philadelphia Association for Critical Thinking).

They had offered $742 000 to any TT practitioner who could reliably pick up a "Human Energy Field" (HEF). They had sent out invitations via email and snail mail to over 60 organisations and individuals involved in TT. Out of 40 000 potential volunteers, only one single well-meaning TT practitioner, Nancy Woods, a California practitioner of Therapeutic Touch, offered to have her abilities tested.

Not Your Regular TT Practitioner

However, Woods was different from most other practitioners of Therapeutic Touch. First, she couldn't detect an HEF from a well person. (But she said that she could pick up an HEF from an unwell person — an injured or painful body part gave her a "*cold, hot or 'pulling' sensation*".) Second, the very act of her trying to detect an HEF would often heal the sick person.

After some discussion with PhACT, she agreed that she should be able to tell the difference between one person with chronic long-term pain in their right wrist, and another person with no medical problems and no pain in their right arm. The person with the chronic long-term pain had not responded to any previous treatments, and the pain would most probably remain for the duration of the experiment. In the experiment, each of the individuals would take their turn at putting their arm into a fibreglass arm sleeve.

Results

Just to make sure that the experimental set-up didn't interfere with Woods' abilities, she performed Therapeutic Touch over the patients' arms while looking at their faces. She said that even through the fibreglass, she could sense the difference between an injured and uninjured arm. She achieved 100% success rate.

But then the experimenters covered the subjects with blankets. Woods could see the fibreglass sleeve, but didn't know whose arm was in it. She couldn't see their faces either.

Suddenly, the results were very different. She could not detect the painful arm. Out of 20 attempts, she was correct 11 times. This success rate is what you would expect from guessing.

What Does This Mean?

This experiment does not prove that human energy fields do not exist. They might exist. One day, with yet-to-be-invented machines, we might be able to reliably detect them. After all, it was only in the 1970s that we invented machines that could detect very powerful bursts of gamma rays coming from deep space.

This experiment also does not prove that practitioners of TT can't detect an HEF. But it does show that this particular practitioner of TT could not reliably tell the difference between HEFs from healthy or unwell people — that is, not without other clues, such as being able to see the patients' faces.

Since then, the prize money has been increased to $US1.1 million for a practitioner who can reliably detect an HEF. Nobody has stepped forward to claim it.

I guess they don't feel lucky.

PLACEBO 3 — MY EXPERIENCE

I became convinced of the power of the placebo very early one morning.

I was a doctor in the **Casualty Department** of a large teaching hospital. A few hours after midnight, a patient whom I knew to have kidney stones came to me with severe kidney pain. His face and his whole body expressed the incredible pain that he was enduring. Patients who have undergone many different types of pain all agree that kidney pain is probably the worst pain that a human can experience. He was already scheduled for surgery in a week, so part of the immediate treatment would be a quick hefty dose of a pain-killing drug. I asked the nurse working with me if she could get some.

Meanwhile, I quickly got him onto a bed, made him comfortable, and got access to a vein. At that moment, the nurse came up behind me and whispered in my ear, *"We've just run out, so I'll go down to the wards, get some more and be back in a minute"*. I nodded, she left, and I was left with my patient in severe pain.

At this stage, I decided to try the placebo effect. I had never tried to invoke it before in my medical career. I kept talking with, and reassuring, my patient, while I drew up some sterile saline water into the syringe. I was hoping that the placebo effect would relieve at least some of his pain, during the unavoidable delay.

I was very careful not to say that I was drawing up the pain-killer. I was also very careful not to say that I was drawing up water. But I deliberately gave him the impression that I was drawing up pain-killer. If he had asked me what I was doing, I would have told him the truth — that I was drawing up saline water. I would also have told him that saline would keep the intravenous line open. But he didn't ask me.

He was in terrible pain, and he kept looking at me as I drew up the sterile saline water, and injected it into his vein.

As I injected it, he began to relax. Within 30 seconds, his face and his body were totally free of any signs of pain. He visibly relaxed and said, *"Thank God, that feels a lot better"*.

That dramatic episode convinced me of the power of the placebo. But the point is this, pain-killers will kill the pain every time, whereas the placebo will do it only some of the time.

It would be very easy for the charismatic, honest and well-meaning healing practitioner to induce the placebo effect — in *some* of their patients *some* of the time.

The Child and the Emperor's Clothes

Since the early 1970s, practitioners of TT have said that they can feel an HEF, assess it, and manipulate it to make it better.

It is the very basis of TT that practitioners can feel the HEF. If they can't feel the HEF, there is no Therapeutic Touch.

Nancy Woods was just one person. Any statistician will tell you that "one" is way too small a number to be a significant sample size.

In 1996, a nine-year-old girl called Emily Rosa decided to test whether 21 practitioners of Therapeutic Touch could "feel" this strange "Human Energy Field". She did this for her fourth grade primary school Science Fair project.

Emily contacted 21 volunteers by placing advertisements in a newspaper. The volunteers had been practising TT for between one and 27 years. The group included nine nurses, seven certified massage therapists, two citizens with no other experience in the healing professions, one chiropractor, one medical assistant and one phlebotomist. Nineteen of them were women. The tests were carried out in two stages — in 1996 and 1997.

Emily designed and carried out the study, and collected the results. However, she did get a few adults (including a statistician) to help her analyse the results, and write a paper. This paper was published in the prestigious *Journal of the American Medical Association*.

How did a nine-year-old girl get published in the *Journal of the American Medical Association*? The editor, George Lundberg said: "*Age doesn't matter. All we care about is good science. This was good science.*"

Experiment 1 — 1996

Emily would put one of her hands near one of the hands of the 15 practitioners. The practitioner would not be able to see which of their hands was closer to Emily's hand.

All the practitioner had to do was say whether their left hand or their right hand was picking up the HEF from Emily's hand.

The practitioners rested their arms on a flat surface, with their palms upward, and about 25–30 cm (10–12 in.) apart. To stop them from being able to see Emily's hand, a screen was placed over their arms.

Do the Experiment

The volunteers then did the first stage of Therapeutic Touch, the "centering" process.

Emily Rosa then flipped a coin to work out which of the practitioner's hands she would test, and then said the word "OK". She put her hand over, say, the practitioner's right hand. At that stage, the practitioner went into the "assessing" stage to work out which one of their hands was picking up the energy field from Emily's right hand. There was no time limit — they could take as long as they wanted. They spent between seven and 19 minutes in this stage.

Each practitioner had more than one trial. Some had 20 trials each, while others had 10.

The Results

If the TT practitioners were just guessing whether Emily's hand was closest to their right or left hand, they would have achieved roughly 50% success rate. And that's what happened.

SIDE EFFECTS OF TT

According to the *New York Times*, Therapeutic Touch can have bad side effects: "*A patient in a Midwestern hospital reportedly complained after a careless Biofield Practitioner, working on someone in the next bed, scooped some negative energy onto him.*"

If the TT practitioners are working with such powerful energies, perhaps they should be more careful.

Shouldn't the excess negative energy removed from the sick patient be treated as a hazardous biological waste? If this energy is real, it certainly should not be thrown in the direction of another patient.

And it shouldn't be thrown out the window either. What if it went out the window down to the local river, and washed up on the coast and a surfer suddenly felt unwell? Who should be sued — the original owner of the energy, or the person who threw it out of the window?

ENERGY FIELDS 1 — NO SUCH THING

The phrase "energy field" is actually meaningless.

Any physicist will tell you that "energy" does not exist in a "field". You cannot have an "energy field". You can have a "magnetic field", an "electric field", and a "gravitational field", but "energy field" means nothing.

ENERGY FIELDS 2 — THE HAND

Arteries carry warm, oxygenated blood into the hands, and veins bring back the cooler, de-oxygenated blood to the heart, and everybody agrees that hands can radiate heat. Heat is a form of electromagnetic radiation.

But no scientist has ever found any other form of electromagnetic radiation coming off the hands, nor from any other part of the human body.

ENERGY FIELDS 3 — THE "SCIENCE"

Proponents of Therapeutic Touch try to explain it scientifically by using "scientific" words. It is almost certain that they don't understand what these words really mean.

They use terms like "Quantum Physics", "electron transfer resonance", "stereochemical similarities", "electrostatic potentials" and "interpersonal energy transfer". They think that if they string enough of these big words together, they can "prove" the existence of "Human Energy Fields".

Their claims have gained wide acceptance by TT practitioners, but little support from the real physicists.

In Emily Rosa's first study, the success rate was very close to what you would expect by chance — 47%. That was 70 successes out of 150 trials.

Problems with the Experiment

This kind of scientific testing was new to both Emily and to the TT practitioners. In science, you do the first experiment so that you can then do a better second experiment. You learn from the first experiment, fine tune it, and incorporate some improvements into the second experiment.

So it's not surprising that the practitioners, once they had a bit of time to reflect on the experiment, offered a few ideas.

Problem 1 — Memory

The first concept put forward by the practitioners was that as Emily Rosa worked her way through the volunteers, she had left a "memory" of her hand behind on the apparatus. Perhaps this effect would have been visible in the results. Perhaps the first practitioners to be tested would have been able to pick up her energy field, but the last practitioners would have had a lot of difficulty.

This is a very reasonable and intelligent suggestion. However, when the results were analysed, the early practitioners had the same failure/success rate as the later practitioners.

Anyhow, if a practitioner can pick the difference between one disease and another, they should be able to pick the difference between a "fresh" field and an "old" one.

Problem 2 — Left and Right Differences

The second complaint was to do with significant differences between the left and right hands in TT. According to some Theories of Therapeutic Touch Practice and Technique, the left hand receives the energy, while the right hand transmits it. This would mean that it would be more difficult for the TT practitioners to detect Emily's HEF with their right (transmitter) hand.

Once again, this is a reasonable suggestion. However, another analysis of the results showed that the failure/success rate was about the same for each hand, and many TT practitioners use both hands to assess the patients.

By the way, there are no experiments carried out by TT researchers that *prove* that "*the left hand receives the energy, while the right hand transmits it*". It has just been stated, and practitioners of TT are expected to accept this statement on faith — not proof.

The practitioners asked that they be able to nominate which of Emily's hands she would offer for testing. Emily was happy to agree.

Problem 3 — "Familiarisation" Time

Thirdly, the Therapeutic Touch practitioners said that they should be given time to identify Emily's HEF before they began the trials.

The TT literature does state that the fields are very different from person to person.

TT TRAINING

Over 100 000 people have gone through TT training. Dolores Krieger claims that she has trained over 48 000 personally.

To get your basic certificate will cost you between $US250 and $US300. The student should practise TT only after they have had a minimum of 12 hours of instruction.

But under one training protocol, the student needs further supervision by a nurse (with a master's degree in nursing and 30 hours of TT theory instruction), and then undergoes 30 hours of supervised TT practice. This all adds up to more money again.

So Emily gave the practitioners time to familiarise themselves with her HEF in the next series of tests.

Problem 4 — Emily to Transmit

The fourth objection to the first test was that Emily herself should have been more active, and centred herself and/or tried to send energy from her own body.

However, one of the fundamental premises of TT is that only the practitioner's intent and centring are necessary. Therapeutic Touch supposedly works because the practitioner detects the patient's HEF. Patients can even be unconscious — they don't have to contribute anything.

Experiment 2 — 1997

Emily agreed to the practitioners' suggestions, and did a follow-up experiment with 13 practitioners, including seven from the first experiment. This time, they had the opportunity beforehand to "feel" and "recognise" Emily's HEF. They could also choose which hand Emily would put near one of their hands.

This time, the test results were even worse. Instead of getting 47% right, they got only 41%. The analysis of the results said this was not really different from guessing.

Does Therapeutic Touch Work?

Therapeutic Touch certainly may "heal" the occasional person, especially if the practitioner is very charismatic. The Placebo Effect can be very effective — at least as a one-off.

ANCIENT QUACKS

Some 2400 years ago, Hippocrates wrote about epilepsy as *the sacred disease.*

He wrote angrily about charlatans or medical fraudsters who deceived their fellow humans. These quacks claimed to be able to cure people of epilepsy. At the same time, they claimed to justify their work by calling upon forces, such as religion, that could not be understood by logic or reason.

"They who first referred (epilepsy) to the gods appear to me to have been just such persons as the conjurors and charlatans now are, who give themselves out for being excessively religious, and as knowing more than other people. By such sayings and doings, they deceived mankind by (performing) lustrations and purifications upon them, while their discourse turns upon the divinity and the godhead."

Things haven't changed much. Today, some "alternative" or "complementary" therapies call upon forces that are not understood by science, or that are at complete variance with what science can prove.

But for reliable healing of patients, who after all are sick and need to be healed, you want something with a better success rate than a placebo.

The essence of Therapeutic Touch is to detect, and then heal, the "Human Energy Field".

If the practitioners of Therapeutic Touch can't pick up the "Energy Field" of a human being, then presumably they can't *manipulate* this mysterious energy either.

And so while Therapeutic Touch definitely has no Touch, it seems that it doesn't have much Therapy either.

REFERENCES

Leon Jaroff, Rita Healy and Jennifer Mattos, "A No-Touch Therapy", *Time*, 21 November 1994, p 88.

Dolores Krieger, Erik Peper and Sonia Ancoli, "Therapeutic Touch — Searching for Evidence of Physiological Change", *American Journal of Nursing*, Vol. 79, No. 4, April 1979, pp 660–662.

Dolores Krieger, "Therapeutic Touch: The Imprimatur of Nursing", *American Journal of Nursing*, Vol. 75, No. 5, May 1975, pp 784–787.

Janet F. Quinn, "One Nurse's Evolution as a Healer", *American Journal of Nursing*, Vol. 79, No. 4, April 1979, pp 662–664.

Linda Rosa, Emily Rosa, Larry Sarner and Stephen Barrett, "A Close Look at Therapeutic Touch", *Journal of the American Medical Association*, 1 April 1998, pp 1005–1010.

"Therapeutic Touch Fails to Detect 'Human Energy Fields' — Further Professional Use of Therapeutic Touch is Unjustified", *Science News Updates*, 1 April 1998.

SUICIDE CELLS – APOPTOSIS

There are about 100 trillion cells in your body. In the time that it takes you to read this little story, several hundred million of your cells will have died. But you won't shrink away to nothing, because these cells get replaced.

Your cells have a self-destruction mechanism. They are always just one step away from activating the mechanism that makes them commit suicide. So, in fact, every cell in your body is permanently teetering on the brink of suicide.

Luckily, most of the time this self-destruction mechanism works only when it's supposed to. Under normal circumstances, it controls the growth and development of the organism, the remodelling of various tissues and organs during life, and the response to various bacteria, viruses and cancers.

When this self-destruction mechanism works too well, your cells die too quickly. You can suffer from diseases such as osteoporosis and rheumatoid arthritis.

When this mechanism does not work well enough, you can get any one of a bunch of cancers, as well as many other diseases.

Cells Must Suicide

Over the last century and a half, there has been an enormous amount of research into the biology of human cells. Your body is made up of many different types of cells — brain, kidney, bone, muscle and heart cells, etc.

Most of this research has been into how cells are born, grow and live. It has looked at the happier side of life.

But only recently have we begun to look at the flipside — how cells die. Why did it take us so long to look at death?

Death is a slightly messy topic. To study death is to realise your own mortality. Like Woody Allen said, *"It's not that I'm afraid to die. I just don't want to be around when it happens."*

Death Is a Part of Life

Some cells have to die so that other cells can live.

You can get a feel for this idea that cells have to die by thinking about how to make an arch from many small tapered stones.

The stones all rest on each other. Once it's standing, the stone arch can stand for thousands of years. A completed stone arch is stable. But you can't build your arch by just piling the stones on top of each other. Once you get to a certain point, the partly completed arch just collapses.

So first, you build a scaffold. Then you build the stone arch on top of the scaffold. Finally, you remove the scaffold, leaving the arch standing triumphantly.

Something like this happens in living creatures.

When a tadpole turns into a frog, the cells on the long, thin tail of the tadpole have to die to make way for the short, stumpy end of the frog. Your hand originally starts off as a little flipper. Cells in that flipper commit suicide in four neat little rows, leaving you with five fingers on each hand and five toes on each foot. In ducks, this process doesn't remove as much of the flipper, so they're left with little webbed feet.

Apoptosis

This process of natural programmed cell death is called "apoptosis". In Greek, *apoptosis* literally means "falling off" or "withering". But in the language of the cells of the human body, apoptosis means the normal and essential process called programmed cell death.

This process of apoptosis removes unwanted cells.

This natural process of programmed cell death happens everywhere in your body.

Apoptosis in Humans

Consider the heart. As a human heart gradually develops, it grows as a solid blob of cells. But along the way, the cells on the inside of the blob die, leaving behind a neatly formed set of four chambers. Apoptosis is essential to make a fully working heart.

In some parts of the developing human brain, 90% of the nerve cells commit a pre-programmed suicide in the right place and in exactly the right order. On average, about 50% of all the cells that started off in your

PRONUNCIATION OF APOPTOSIS

It's confusing, but there are two ways to pronounce "apoptosis".

The word "apoptosis" comes from the Greek words *apo* (meaning "by") and *ptosis* (meaning "to drop" or "fall"). *Ptosis* by itself is pronounced "TOE-sis". A medical doctor will refer to the "ptosis of an eyelid", meaning that one eyelid droops lower than the other eyelid.

One school of thought loves the Classical Greek language. They say we should pronounce each Greek word separately, so you don't sound the second "p". This means that "apoptosis" gets pronounced "AP-oh-TOE-sis". For example, *Steadman's Medical Dictionary* leaves out the second "p".

But another school of thought says that we can ignore the classical roots of a word. Instead, we should pronounce a word so that it sounds "right". This school has won. The general rule seems to be that we do pronounce the second "p" — "AP-pop-TOE-sis". *Dorland's Medical Dictionary* follows this rule.

But sometimes things are easier. Both Dorland's and Steadman's dictionaries pronounce the second "p" in "proptosis".

nervous system when you were an embryo will die before you are born.

During menstruation, a woman bleeds because the cells that lined the uterus have gone through their natural programmed cell death.

When the cells that squirt dye into your hair begin to commit suicide, you end up with the white hair of old age.

Your immune system knows the difference between you and the rest of the universe. Whenever something foreign enters your body, cells from your immune system will wrap themselves around it. Then these immune cells will commit suicide in apoptosis, annihilating the invader at the same time. This is one of the mechanisms that your immune system uses to kill bacteria and viruses.

But occasionally, some cells in your immune system get a bit overexcited and try to attack other cells in your body. These "rogue" or overactive immune cells get hunted down and told to commit suicide.

Our Cousin, the Worm

We have learnt a lot about apoptosis by looking at one of the Top Ten favourite laboratory animals, the nematode worm, *Caenorhabditis elegans*.

This worm has been popular for a long time, but it became really famous in late 1998. It was then that it became the first animal to have its complete DNA decoded. (Your human DNA is a very long molecule, a few metres in length. It looks like a ladder, with two side rails and about three billion "rungs".) The worm's DNA has only 97 million "rungs". It took 15 years, $US60 million, and 1500 scientists in 250 laboratories across the world to map all of this worm's DNA.

Even though it's only 1 mm long and has only 959 cells, about 40% of the worm's genes are closely related to ours. Just like humans, this worm has sex, has a nervous system, moves by using muscles, eats food and has a digestive system.

To make one of these tiny little worms, the worm's DNA grows exactly 1090 cells. But not all of those cells survive to be a wriggling worm. Before the worm finishes growing, exactly 131 of those 1090 cells have to die in exactly the right place and at exactly the right time. One hundred and thirty-one of these cells have been born only to die.

This process of apoptosis is so tightly controlled that some of these suicidal cells have a total life span of less than 60 minutes.

Two Ways to Die

There are two main ways in which cells can die: the accidental death called "necrosis", from the Greek for "make dead", and the programmed death called "apoptosis".

Necrosis is an extremely messy event. A cell will die by necrosis after being treated badly, such as being poisoned, overheated, or starved of oxygen. As the damaged cell heads towards its inevitable necrotic death, it will swell, and its cell membrane will split open. The chemicals inside leak out and cause damage and inflammation to the other innocent cells around them. So all the cells around the dying cell will suffer.

But cell death by apoptosis is a much tidier way to die. The cell will shrink in a very neat way. The bits left over will be neatly mopped away, and life will continue as normal in the neighbouring cells.

Necrosis and Apoptosis Together

Sometimes, necrosis and apoptosis are all jumbled up together.

In a stroke, the blood supply to part of the brain is cut. In a heart attack, the blood supply to part of the heart is cut.

At first, the cells die by necrosis. They swell and burst open. The contents of these cells can damage the cells nearby. But then the first wave of destruction slows. A second wave of death spreads outwards, via apoptosis.

We don't yet know how this wave of apoptosis is unleashed. Once we do, we may be able to stop it. If we can stop it, we can limit the damage caused by a heart attack or a stroke.

Stages of Apoptosis

There are four definite stages in the programmed death, or apoptosis, of a cell: the decision to die; the actual carrying out of the death sentence; the engulfment of the tiny fragments of the cell; and the final degradation and vanishing of the cell.

First, the condemned cell follows instructions from outside, or from its own DNA, to kill itself. The doomed cell makes the chemicals that it needs to kill itself, and then it unleashes these chemicals upon itself.

In the second stage, these chemicals go to work. One of these chemicals (a nuclease) slices up the DNA into thousands of tiny fragments. The other chemical (a transglutaminase) makes the doomed cell change shape. It does this by joining various tiny biological machines in the cell together with chemical bonds. As it changes shape, the suicidal cell gradually shrinks and splits, and re-packages itself into smaller fragments. At all times, these fragments stay wrapped up inside a cell membrane, so there's no leaking of chemicals into the immediate environment, and none of the nearby cells are affected.

In the third stage, cells from the immune system come along and gobble up the fragments.

In the fourth stage, the immune system cells begin to degrade the tiny fragments. When they are finished, the immune cells leave, all signs that the cell ever existed vanish, and the area is all neat again.

Now here's an odd fact. The cell has to be in good condition before it can make the "killer" chemicals. In other words, it has to be healthy before it can commit suicide. Some "unwell" cells cannot kill themselves.

History of Apoptosis

Apoptosis really took off as a research topic in the early 1980s. However, various scientists had skirted around apoptosis for one-and-a-half centuries, without really coming to grips with it.

One of the first discussions of programmed cell death was written way back in 1842. Carl Vogt described how when nerve and bone cells go through their normal development, some of the cells died naturally.

In 1864, Weismann looked at the cells of pupating diptera as they developed. He saw massive amounts of programmed cell death. In 1872, Stieda looked closely at bone as it developed. He wrote about how cartilage cells had to die so that bone cells could grow. In 1883, Metschnikoff wrote about how when tadpoles began to turn into frogs, immune cells came along and ate the fragments of the other cells that had died. In 1889, Beard wrote about how as a fish embryo grew, huge populations of neurons simply vanished.

Around the end of the 19th century, between 1893 and 1913, Barfurth wrote papers every year about this concept of programmed cell death, which he called "involution".

In 1889, Felix described how some cells would die, and others would survive, in what would eventually turn into the

WARNING: APOPTOSIS AND "DISEASE"

In the following boxes, I have given a bunch of examples of how apoptosis affects several different diseases.

These examples are extremely simplified. Even so, they are still very complicated. You will read how "this chemical" acts on "that chemical" which is bound to "another chemical" which, after another 50 steps of "other chemicals", can cause (or prevent) the disease. Another problem is that the chemicals all have really weird names.

Don't blame me.

The Human Body (THB) is really very complicated.

I would guess that it's more complicated than the Beginning of the Universe (TBOTU) or Quantum Physics (QP). But it's hard to tell, because we don't yet have the Whole Story on THB, TBOTU or QP.

APOPTOSIS AND LUNG CANCER

Adi Kimchi and colleagues from the Weizmann Institute of Science in Rehovot, Israel, found that apoptosis is involved in controlling one of the four main lung cancers.

An enzyme called DAP kinase helps trigger apoptosis. But Kimchi and the team found that their lung cancer cells did not make this enzyme. No DAP kinase meant that their lung cancer cells would not commit suicide. The cancer cells would live and grow bigger.

But then they restored DAP kinase to cancer cells. The cancer cells slowed down and grew more slowly.

The DAP kinase also had another good effect. Some cancers make little groups of cancer cells. These tiny blobs can float through the bloodstream and land somewhere else in the body (such as the bone) and set up another cancer colony. These are called "secondary" cancers, or "secondaries". The DAP kinase stopped the main group of cancer cells from making secondaries.

But this research is just laboratory research so far. It has not yet been tried on people with lung cancer.

muscle of a mammal. In 1906, Collin wrote about his discovery that as nerve cells grew, some of them had to die. He actually argued that you had to make too many cells in the first place, so you could successfully cull some away.

After a gap of half a century, in 1951, Glücksmann also recognised that some cells had to die as a part of normal development.

In 1959, F. Macfarlane Burnet thought about cells in the immune system that could kill healthy cells in your body. He proposed that these cells with "potential self-reactivity" died while still in the embryo. This was part of his Nobel prize–winning work on the Clonal Selection Theory of Acquired Immunity.

Many other scientists also saw and described this apoptosis. But they didn't understand the importance of what they were seeing. Most cell biologists were more interested in understanding the life of the cell, not its death.

Apoptosis really became a hot topic about 30 years ago, thanks to the Australian John Kerr. In 1995, Professor John Kerr retired from being head of the School of Pathology at the University of Queensland. But in the early 1970s, he was studying under another Australian, Sir Roy Cameron, at the University of London. He was given the job of looking at what happened to cells when they were damaged and then died.

As he expected, he saw many cases of the standard and well-known cell necrosis (where the dying cell bursts and dumps its guts all over the place). But he also saw a few cases of some cells that shrank "*as though someone has tightened a net around* the cell, shrinking it down". He called this strange phenomenon "shrinkage necrosis", and wrote about it in the *Journal of Pathology* in 1971 and 1972.

In 1972, Kerr teamed up with the British pathologist Alistair Currie and Currie's Ph.D. student Andrew Wyllie, and wrote a paper in the *British Journal of Cancer* called "Apoptosis: A Basic Biological Phenomenon with Wide-Ranging Implications in Tissue Kinetics".

They were spot on. Today, we know that various diseases are linked to too much apoptosis or too little.

APOPTOSIS AND P53 — 1

When our cells turn cancerous, they usually kill themselves via apoptosis. This is because, luckily for us, our DNA has many anti-cancer genes built into it.

One of the most famous anti-cancer genes is called p53. It's a Master Switch between life and death. It's called a "tumour suppressor" gene. The p53 gene is called "53" because it makes a protein that has a molecular weight of 53 000. This protective gene is also called the Guardian of the Genome.

Once the cell is damaged by cancer, radiation, or diseases, p53 triggers the cell to commit suicide by apoptosis. The damage or disease is stopped.

But when p53 is damaged, cancer cells can grow rapidly. If p53 is deactivated in mice, they will die of cancer within three months.

The human papillomavirus tends to inhibit and weaken p53. Papillomavirus is a major cause of cervical cancer.

If you look at human cancers in general, you will find a deactivated p53 gene in over half of them. The p53 gene was, in most cases, deactivated by a mutation.

APOPTOSIS AND P53 — 2

Bert Vogelstein and colleagues from the Johns Hopkins Oncology Center and the Johns Hopkins University School of Medicine in Baltimore, Maryland, found one of the pathways by which p53 set off apoptosis.

When it is stimulated by being exposed to a cancer, the p53 gene will activate tens of other genes. Many of these will make a whole bunch of different toxins called Reactive Oxygen Species (ROS). The ROS toxins then destroy the mitochondria in the cancer cells. The mitochondria make energy — and without energy, the cancer cells die.

Too Much Apoptosis

First, let's look at some diseases where you have *too much apoptosis*.

These include AIDS, where some of the central cells of the immune system die far more rapidly than they should; and various neurodegenerative disorders, such as amyotrophic lateral sclerosis and retinitis pigmentosa. Other groups of illnesses include blood diseases, such as aplastic anaemia; low-oxygen injury, such as heart attack, or stroke; or even toxin-induced liver disease, such as caused by alcohol.

Things that will stimulate apoptosis include natural chemicals, such as glutamate, dopamine, calcium and glucocorticoids; damaging factors, like heat shock, infection by viruses or bacteria, or free radicals; heavy-duty drugs, such as anti-cancer drugs like cisplatin, vincristine and methotrexate; various radiations, such as gamma or UV; and, of course, toxins such as alcohol.

Too Little Apoptosis

But on the other side of the coin, there's a whole bunch of diseases that happen when you have *too little apoptosis*.

You can get a range of different cancers, including breast cancer, prostate cancer, and follicular lymphomas. Too little apoptosis can also cause auto-immune diseases, such as lupus and glomerulonephritis, and a whole family of viral infections, such as those caused by the herpes family, the adenovirus family, and so on.

Today, we know a very small number of the many factors that will slow down apoptosis. These include natural physiological inhibitors, such as zinc, oestrogen, androgens, and various growth factors;

viral genes, such as in adenovirus, cowpox virus, and Epstein-Barr virus; and various pharmacological or drug agents, such as phenobarbital.

Just Right Apoptosis

Apoptosis sculpts many of our body parts.

In our eye, the lens helps focus the incoming light. The lens is made from apoptotic cells that have dissolved their usual internal cell stuff and replaced it with a perfectly transparent protein called crystallin.

You have about 10 metres (or yards) of gut. The cells lining the inside are continually migrating up from the wall to the inner surface. Once they get there, they die and are replaced by more cells.

Your skin is your largest organ — about 15% of your body weight. Skin cells start off in the deeper layers, and then, over the

APOPTOSIS AND CANCER — SURVIVIN

Dario C. Altieri and colleagues from the Yale University School of Medicine looked at a chemical called survivin.

Survivin is a protein that inhibits apoptosis. If too much survivin is made, there won't be much apoptosis happening. Not enough apoptosis can trigger some cancers.

APOPTOSIS AND FAS — 1

The protein called Fas is very famous in the literature of apoptosis. It's one of the triggers that sets off cell suicide. Scientists call Fas the "death protein" or "death receptor". It's a receptor protein that sits on the outside surface of the cells in our body.

Fas has a "partner" — FasL. FasL is a different protein that sits on the outside of activated killer T-cells (part of your immune system).

Fas and FasL are a perfect match for each other. They are like a lock and key. FasL is the "key" that slips inside the "lock" that is Fas. Under normal conditions, this does not happen.

Fas changes when your normal cells get infected or cancerous, and it will attract FasL. As soon as FasL enters Fas, it sets off the rapid self-destruction of the cell that carries the Fas.

APOPTOSIS AND FAS — 2

Even the killer T-cell gets killed.

The T-cells that carry FasL live for only a few days. When they are ready to die, they will make their own Fas on their outside surface. A younger T-cell will come along, insert the FasL protein, and tell the old T-cell to die.

APOPTOSIS AND FAS — 3

The process of apoptosis can kill off cells that are infected with a virus, or that have turned into cancer cells. Fas helps look after this particular application of apoptosis.

Avi Ashkenazi and colleagues from Genentech Inc. in South San Francisco, California, looked at a chemical called decoy receptor 3 (DcR3). This decoy receptor chemical will stick to some of the chemicals involved in Fas, and stop them from working. When Fas doesn't do its job, apoptosis doesn't get triggered.

So, if a bad (virus-infected or cancer) cell can make DcR3, it can escape attack from the immune system, and apoptosis.

The team discovered that DcR3 was found in about half of the 35 primary lung and colon cancers that they studied.

the stages of APOPTOSIS

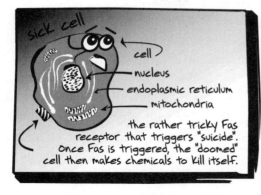

the rather tricky Fas receptor that triggers "suicide". Once Fas is triggered, the "doomed" cell then makes chemicals to kill itself.

- cell
- nucleus
- endoplasmic reticulum
- mitochondria

sick cell

1. the "condemned cell"

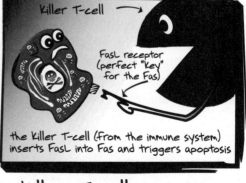

Killer T-cell

FasL receptor (perfect "key" for the Fas)

the Killer T-cell (from the immune system) inserts FasL into Fas and triggers apoptosis

2. Killer T-cell arrives...

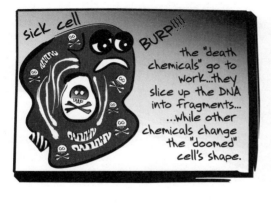

sick cell

BURP!!!!

the "death chemicals" go to work...they slice up the DNA into fragments... ...while other chemicals change the "doomed" cell's shape.

3. chemicals doing their thing...

Killer T-cell

condemned cell fragments

score
Killer T-cell : 1
cell : 0

4. let's eat...

next 21 days, gradually migrate to the surface. They die during this journey. They stay on the surface as a protective layer. If the skin cells are damaged on the way up, for example, by too much sunlight, they will commit suicide early.

Control Apoptosis?

You can see that if you have too much or too little apoptosis, you can get a whole bunch of different diseases. And that is part of the reason for the incredible interest in apoptosis today. If we can learn how to control apoptosis, we can then start to control some of these diseases instead of just treating them.

In Paris, in 1925, Antonin Artaud, the French theatre producer, actor and theorist wrote: *"If I commit suicide, it will not be to destroy myself, but to put myself back together again"*. Suicide for humans is a terrible and destructive act. But suicide for cells is part of living.

Breast Milk Protects Against Cancer?

Research into apoptosis is giving us some really weird results. One scientist has even found that, thanks to apoptosis, human breast milk actually has an anti-cancer effect!

This is a good example of "serendipity" or plain old good luck. The scientists were looking for one thing, but they accidentally made a much bigger discovery.

First, Breastfeeding and Infection

This work was done by Anders Håkansson and colleagues from Lund University in Sweden. Originally, they were trying to work out exactly how breastfeeding protects babies from infections of the gut and lungs. They had a reasonable theory.

They figured that bacteria and viruses could attack you only if they could stick to the lining of your gut or lungs. If they couldn't stick, the bacteria and viruses would get washed out of the baby's bottom, or blown out of its lungs, without causing any harm. So they guessed that in breast milk, there might be something that stopped bacteria and viruses from sticking to the lining of the gut and lungs.

The Swedish scientists then started experimenting with adenovirus (which can cause the common cold), breast milk, and human cells. But their adenovirus took 48 hours to do its damage. Normal human cells were too fragile in the laboratory, and died from other causes long before the 48 hours were up.

They needed human cells that were tougher and that would live longer.

Then, the Lucky Choice

The scientists then made a lucky decision.

They decided to continue their research with human lung cancer cells. These cells are virtually immortal. They also switched

over to bacteria. They set up a glass dish with breast milk, bacteria and some of the young human lung cancer cells.

To their surprise, within half an hour, the cancer cells were all dead. Apoptosis, or programmed cell death, had killed them.

Change of Path

The Swedish scientists then dropped their research into infections and looked at cancers. They found that a common protein in human breast milk, called alpha-lactalbumin, tells young cancer cells to commit suicide. It's a trigger for apoptosis. Once the cancer cells have been told to die, they obey almost immediately.

This looked like a way to attack some of the lung cancers.

But this research came to a dead end. Unfortunately, this protein does NOT kill mature grown-up cancer cells. It attacks only immature cancer cells.

Even so, this is a very interesting piece of research. There are about 1000 different cancers with many different causes. We think that about one third of all cancers

happen because the cancer cells manage to avoid apoptosis. They have some as yet unknown trick that lets them slip out of obeying the instruction to kill themselves.

If we understand apoptosis better, we will also understand cancers better — and maybe learn the command that tells cancer cells to commit suicide.

So, besides being a part of natural growth, apoptosis also protects us from various infections and cancers.

Evolution of Apoptosis

We know there are lots of advantages to having our cells programmed to die, but how did apoptosis evolve?

Frankly, we don't know.

One theory is that it might have allowed the fittest cells in the colony to survive, while the not-so-fit cells were quietly told to go off and die.

Another evolutionary advantage of apoptosis is that it helps the living creature end up with the right amount of cells for the environment. If it has too few cells, it can't fight off invaders. If it has too many

APOPTOSIS AND ASPIRIN

Why do anti-inflammatory drugs such as aspirin and ibuprofen seem to be able to ward off colon cancer?

Yep, the answer is apoptosis again — according to Patrice J. Morin from the Howard Hughes Medical Institute at Johns Hopkins Medical Institutions in Baltimore.

The cells in the colon constantly renew themselves. As they age, they migrate towards the inner surface of the colon. In general, only older cells turn cancerous. But as the cells get near the surface, they usually commit suicide.

The non-steroidal anti-inflammatory drugs such as aspirin and ibuprofen stimulate colon cells into apoptosis. These drugs stop the cells from getting too old and possibly turning cancerous.

cells, and it's living in a harsh environment, there might be too many cells to feed, and it might starve to death.

In our evolutionary history, there's a whole lot of apoptosis going on. In fact, scientists have found apoptosis in various primitive creatures that go back some 2000 million years — nearly half the age of our planet.

So apoptosis is not a fate worse than death, it's a very necessary part of living, where one healthy life depends upon millions of well-timed cell deaths.

The Real Meaning of Apoptosis

Apoptosis is suicide without sadness.

In 1986, Václav Havel, the Czech play-wright, who was also the president of Czechoslovakia, and who awarded high state honours to the rock musician Frank Zappa, wrote about suicide. It was almost as though he was thinking about apoptosis. He wrote: "*Sometimes I wonder if suicides aren't in fact sad guardians of the meaning of life.*"

In the classic movie *The Third Man*, Trevor Howard advises Joseph Cotton to "*leave death to the professionals*". He also wisely says that "*death's at the bottom of everything*".

Many scientists think that the 21st century will be the Century of Life (as opposed to Electronics, or Physics).

The real goal behind all the research is to be able to control apoptosis, perhaps with drugs. We want to be able to activate apoptosis in some diseases and slow it down in others. We still don't understand the process. If we can understand apoptosis, medical science will be reborn.

APOPTOSIS AND OPIUM

Chemicals from the opium poppy have been used for thousands of years to relieve pain. When they're used wisely for pain relief, they're amazingly effective and have very few negative side effects.

But it has been noticed that opiates can affect the immune system.

Yufang Shi and colleagues from the Jerome H. Holland Laboratory for Biomedical Sciences at American Red Cross in Rockville, Maryland, looked at morphine. They found that morphine will stimulate a lymphocyte to make the so-called "death protein" Fas on its surface. A lymphocyte is one of the important cells of the immune system. So Fas can trigger cell suicide in some of the cells of the immune system.

The scientists wrote "*The discovery that opioids induce Fas expression not only bears upon the consequences of taking morphine, but should help our understanding of the interaction between the nervous and immune systems*".

We don't yet know what subtle effects morphine may have on apoptosis. It might even be that apoptosis may cause addiction! But at the moment, we don't know enough about apoptosis.

PLANTS AND APOPTOSIS

You might think that we humans have a hard time surviving. But the plants have it much harder. If somebody comes out to get them, they can't run away. If the weather goes bad, they can't hide under shelter. Their roots grow in soil that is loaded with bacteria, viruses and fungi. The air is full of the same germs, trying to attack them from above.

So how do plants manage to survive?

The important thing to realise is that plants do not have a circulatory system like ours. So they need a few different types of protection.

First, their waxy surfaces and cellulose cell walls provide a physical barrier to invaders. These surfaces are reinforced with various chemicals (such as antibiotics) to fight off the invaders.

As soon as the plant's cell membranes recognise an invader, they release free radicals, which attack the invaders. The free radicals will cross-link with chemicals called lignin. This makes the cell walls resistant to attack from chemicals released by the invaders.

The plants also release quite small antibiotic chemicals called phytoalexins ("plant defenders"). These deter bacteria.

Reactive Oxygen Species (ROS) will cross-link chemicals in the cell wall called glycoproteins, again making the cell walls more resistant to enzymes from the invader.

But the plant has a second line of defence.

Once an invader gets through, the cells of the plants will go to any extreme to protect themselves — including committing suicide for the common good.

The first set of cells that get broken into by an invader commit suicide. They protect the rest of the plant by locking the invading germs into a ball of kamikaze cells.

Cells that are right next to the invader will collapse. This action will further help wall off the invader.

And a little bit further away, plant cells will grow little balls inside them. These balls are full off various chemicals, some of which are antibiotics. If the invader gets near these cells, the little balls rupture open, releasing a whole flood of really nasty chemicals.

Within 12 hours of first contact by the invader, up to 70% of the affected plant cells will be dead, but they will have taken the invader with them.

We don't know too much about plants yet, but it seems that the cell death that happens in plants is very similar to the apoptosis that happens in our cells. The botanists called the plant version of apoptosis Hypersensitive Cell Death.

But apoptosis in plants is, just as in animals, also necessary for normal growth and development. It also causes the leaves to wither as winter approaches.

APOPTOSIS AND LUNG CANCER

In 1995, 150 000 Americans died of lung cancer. In that same year, 170 000 new cases were diagnosed. Lung cancer is mostly caused by cigarette smoking. While men have begun to heed the warning not to smoke, young women have not. Lung cancer has been declining in men but increasing in women. In fact, over the last 10 years, lung cancer is the major cancer in America that kills women.

There are several types of lung cancer. One of them is called small cell lung cancer (SCLC). This cancer is an especially nasty lung cancer to have, because the cells are very resistant to the drugs used in chemotherapy.

Tariq Sethi and colleagues from the University of Edinburgh Medical School might have found why.

SCLC cells are wrapped in a mushy extensive Extracellular Matrix (ECM). They're joined to this ECM via a chemical called beta-1-integrin. When beta-1-integrin is joined to the ECM, the chemotherapy drugs can't trigger apoptosis in the cancer cells.

The scientists think that if they can find ways to block the action of beta–1-integrin, they may be able to find a way to make the SCLC cells sensitive to the drugs of chemotherapy.

REFERENCES

Richard C. Duke, David M Ojcius and John Ding-E Young, "Cell Suicide in Health and Disease", *Scientific American*, December 1996, pp 80–87.

Martin Raff, "Cell Suicide for Beginners", *Nature*, Vol. 396, 12 November 1998, pp 119–122.

Martin C. Raff, "Death Wish", *The Sciences*, July–August 1996, pp 36–40.

Tariq Sethi et. al., "Extracellular Matrix Proteins Protect Small Cell Lung Cancer Cells Against Apoptosis: A Mechanism for Small Cell Lung Cancer Growth and Drug Resistance in Vivo", *Nature Medicine*, June 1999, pp 662–668.

Hermann Steller, "Mechanisms and Genes of Cellular Suicide", *Science*, Vol. 267, 10 March 1995, pp 1445–1462.

Stephen Young, "Life and Death in the Condemned Cell", *New Scientist*, No. 1805, 25 January 1992, pp 22–25.

SEX AND WATER ALLERGIES

People can be allergic to all sorts of substances — plant pollen, the droppings of the common dust mite, various pharmaceutical drugs, nickel, and even various chemicals in some wooden furniture.

But if you go looking in the medical literature, you'll find even rarer allergies, such as allergies to water, and even allergies to the male sexual ejaculate, semen.

Allergic Reaction

Your immune system has the job of defending your body from invasion by bacteria and viruses, etc. Most of the time, your immune system is pretty sensible. It will mount a little immune response against a little invader and a big immune response against a powerful invader.

An "allergic reaction" happens when a part of your immune system overreacts to something that it doesn't really need to react against, such as the usually harmless pollen of the flowers in springtime.

The allergic reaction of the unfortunate sufferer can vary enormously.

Sometimes, it can be a mild skin rash with a little bit of local itching. Sometimes, it will progress past this to swelling around the eyes, sneezing, blocked nose, vomiting and diarrhoea. In extreme cases, the allergic reaction can make the airways swell so much that you're in severe danger of dying from lack of air. In these cases, the allergic reaction is called "anaphylaxis" — from *ana* meaning "wrong" and *phylaxis* meaning "to guard".

Allergic to Sex

In the Swinging 60s, sex was "safe". But nowadays, much to Austin Powers' distress, it's not. The semen can carry germs.

But what's so dangerous about germ-free semen, apart from the risk of getting pregnant?

The semen has an average volume of around 2.5–3 ml (about $\frac{1}{10}$ fluid oz) after several days of abstinence.

The semen has two parts — millions of sperm, and the liquid that they swim in. This liquid is made up of secretions from various glands, such as the prostate gland, Cowper's glands, the seminal vesicles, and probably the urethral glands.

Over the years, we've been looking for the chemicals that set off this allergic reaction. They appear to be in the liquid part rather than the sperm part. At the moment, the finger of suspicion is pointed at a bunch of chemicals that have a molecular weight somewhere between 15 000 and 30 000. This family of molecules seems to come from the prostate gland. But we still haven't isolated the dangerous molecules yet.

History of Sex Allergy

First of all, this is a very rare reaction. After all, if a woman is allergic to semen, she probably won't have children, which makes it hard to pass on this allergy. However, Dr Jonathan A. Bernstein from the Division of Immunology at the University of Cincinnati College of Medicine thinks that this allergy might be more common. He thinks that there may be many women with very "mild" symptoms, who may not recognise or complain about them.

The very first case of an allergic reaction to human semen was reported by Specken back in 1958. Since then, we've probably had less than 100 cases reported in the medical literature. The official name of this allergy is Human Seminal Plasma Hypersensitivity.

The average sufferer in these reports was a 25-year-old woman who noticed that immediately after a close encounter with semen, she had a generalised itchiness over her whole body. But when her partner used a condom, she had absolutely no symptoms at all. She had a family history of some kind of allergic disorder, plus a personal history of rhinitis or asthma, and perhaps an allergy to some foods.

MALE EJACULATE — SPERM

Sperm are made in the testicles, on the walls of the seminiferous tubules. It takes about 74 days to turn a primitive germ cell into a mature sperm.

The testicles are held away from the body, so they are kept at a lower temperature — typically, about 32°C (90°F). When men increase the average temperature of their testicles, by having long hot baths and wearing tight athletic supporters that keep the testicles closer to the body, the sperm count drops a little, but not enough to make "hot-baths-and-athletic-supporters" a reliable male contraceptive.

Even though you need only one sperm to fertilise the egg, there are around 100 million of them in each cubic centimetre of semen. Once they've been delivered, the sperm move at around 3 mm (0.11 in.) per minute through the female genital tract. They reach the fallopian tubes, where fertilisation usually happens, about 30 to 60 minutes after ejaculation.

In a few reports, the woman's allergy to semen began after either she or her male sexual partner had had some kind of surgery to the genital tract. This could include a man having a partial removal of his prostate or a vasectomy, or a woman having her uterus removed.

Symptoms of Sex Allergy

The symptoms can be either local to the genital tract, or generalised over the whole body, or both. They can range from mild to life-threatening.

So, within five to 10 minutes after sexual intercourse, the woman begins to experience swelling around the vagina, sometimes with the sensation of obstruction in the nose, severe sneezing, swelling around the eyes, and generalised itchiness over the whole body. In extreme cases, she could have the frightening sensation of her throat swelling, causing difficulty in breathing.

Luckily, the severe reaction is quite rare.

It's actually more common to have the symptoms develop over a longer period of time, and for them to be less severe. Some patients reported that the symptoms gradually worsened over a few days, and then spontaneously resolved by themselves.

Cause and Cure

One study came up with a very unusual finding. They found that in all of the couples they tested, the man and the woman each had very similar immune systems. This can sometimes happen when women choose their partners while they are taking the Pill.

The Pill will change a woman's sense of smell, so that she is attracted to men who smell the same as she does. This similar smell is linked to having a very similar immune system, which can be a problem. (See the story "Sex, Smell & Separation" in *Pigeon Poo, the Universe and Car Paint* for more details.)

This semen allergy can cause a huge amount of anxiety and stress — simply because the couple can no longer have free and spontaneous sex. Of course, it can also interfere with their ability to start a family.

Some patients get considerable relief with "selective desensitisation immunotherapy". For example, the woman might have injections of purified proteins from the

MALE EJACULATE — SEMINAL FLUID

The seminal vesicles pitch in about 60% of the total volume of the seminal fluid. They dump in chemicals such as fructose, ascorbic acid, prostaglandins, phosphorylcholine, ergothioneine, and flavins.

The prostate gland contributes around 20% of the total volume, and it dumps in chemicals such as spermine, citric acid, cholesterol, phospholipids, fibrinolysin, fibrinogenase, zinc and acid phosphatase.

And, of course, there are probably a squillion other chemicals that we don't yet know about.

semen. The sooner that this treatment is started, the better it will work. A condom will also stop the symptoms. Of course, you could always do a Cliff Richard, and become a Born Again Celibate.

Allergic to Water

But there's an even rarer allergy than the one to semen — an allergy to water. The medical name for this is "aquagenic urticaria". There are only about 30 of these cases in the medical literature. The youngest sufferer on record is only three years old.

A typical recent case is that of a 20-year-old Vietnamese man living in California. Since the age of 10, if water touched his skin, he would react within five minutes. He had severe itching, red skin with white welts or lumps, headaches and extreme difficulty in breathing. The reaction lasted about 20 minutes. It happened with any kind of water (tap water, sea water, pond water etc). It occurred whether the water was hot or cold.

Water is an essential part of the body; it makes up about 70% of our body weight. It's almost impossible to avoid touching water with your skin during the day. But our young Vietnamese man had to learn very quickly how to wash and dry himself very rapidly indeed.

In his case, we've got absolutely no idea what's going on. The reaction didn't seem to be related to the chemical called histamine, which is very often part of an allergic reaction. He had had his histamine levels checked during a reaction, and they didn't change at all. The doctors tried treating him with the usual anti-allergy drugs (antihistamines and anticholinergics), but they made absolutely no difference at all to his reaction.

Allergic to Sex and Water?

It's true that these are very rare allergies. But if you were really unlucky, and allergic to both water and semen, you couldn't get either clean or dirty.

Jonathan A. Bernstein et. al., "Prevalence of Human Seminal Plasma Hypersensitivity Among Symptomatic Women", *Annals of Allergy, Asthma and Immunology*, Vol. 78, January 1997, pp 54–58.

Kathleen Fackelmann, "Sex Allergy: No Laughing Matter", *Science News*, Vol. 151, 22 February 1997, p 124.

Bernard B. Levine, Reuben P. Siragania and Isaac Schenkein, "Allergy to Human Seminal Plasma", *New England Journal of Medicine*, Vol. 288, No. 17, 26 April 1973, pp 894–896.

K.V. Luong and L.T. Nguyen, "Aquagenic Urticaria: Report of a Case and Review of the Literature.", *Annals of Allergy, Asthma and Immunology*, Vol. 80, No. 6, June 1998, pp 483–485.

M. Medeiros Jr., "Aquagenic Urticaria", *Journal of Investigational Allergology & Clinical Immunology*, Vol. 6, January–February 1996, pp 63–64.

REFERENCES

OLD EARS BIGGER?

There are many burning questions for humanity, and usually the most important ones get answered first. That makes it hard to understand why it took so long to answer the question "Why do your ears get bigger as you get older?".

Anatomy of the Ear

There are three parts to the ear — the outer ear, the middle ear and the inner ear.

The auricle is the external bit of your outer ear that holds up your glasses. It helps collect sound from down in front of you, and off to the side. This would have helped our hunter-gather forebears pinpoint most of the suspicious forest noises.

The outer ear reaches all the way in to your eardrum. Sound enters the outer ear and is slightly amplified by the tapering hole. The pressure waves, from the sound, push on the eardrum. The eardrum is a very sensitive device. It has been claimed that when you listen to the softest sound that you can possibly hear, your eardrum moves back and forth over a distance equal to $\frac{1}{10}$ of the diameter of a hydrogen atom.

Next is the middle ear. The middle ear has three tiny bones. When the eardrum vibrates, the three bones move and they push on a flexible window. There are various kinds of joints between the bones. Synovial joints are the Rolls Royce of joints — they have the lowest friction. These three bones in the middle ear have the smallest synovial joints in the body.

Finally, there is the inner ear. It's full of an electrically charged fluid. When the flexible window moves, pressure waves are set up in this fluid. The pressure waves bend the "hairs" on hair cells. When they bend, they generate electricity which goes via the eighth cranial nerve to your brain — and suddenly you have the impression of sound.

Eastern History of the Long Ear

The Chinese have placed great significance on the length of the ear and earlobe for centuries. Today, Chinese physiognomists declare that large ears show the need "*to draw on inner reserves of strength and ability*". ("Physiognomy" is the skill of working out a person's character from the features of their face or the shape of their body.)

The historical Chinese belief is that the longer your earlobe, the longer you will live. Kay-Tee Khaw, Professor of Clinical Gerontology at Addenbrooke's Hospital, Cambridge, wrote about "*my grandmother (one of the last generation of Chinese women with bound feet), whom I remember admonishing me in early childhood to stretch my ears daily to ensure long life*".

The Chinese also believe that the thicker the earlobe, the thicker your wallet.

It is also said that the longer the whole external ear, the wiser and more noble you are. This might be why Buddha and the emperors of Old China are said to have had long ears. Hindus admire long ears, because of the long ears of the elephant god, Ganesha.

In fact, Liu Bei, who started the Han Dynasty in 221 AD, apparently had ears so long that they reached down to his shoulders. He could supposedly see his ears just by turning his head!

Western History of the Long Ear

But Europeans have a different culture. They tease people who have long ears using names such as "bat ears", "wing nut" and "Dumbo".

Hamish and Laing wrote about this in their paper, "Prominent Ears: A European Perspective" in the *British Medical Journal*. They wrote: "*Many languages identify the potential aerodynamic effects of ear*

prominence . . . Other languages compare (people's ears) with animals' ears: donkey ears in Hungarian and pigs' ears in Polish."

They thoughtfully compiled a list of European insults that relate to long or prominent ears. The list included "teapot ears" (Austrian German), "flapping ears" (Dutch), "flying ears" (Norwegian), "cup handle ears" and "windy ears" (Spanish), and "sail ears" and "ladle ears" (Turkish).

Modern Science Looks at the Ear

But as far as medical science is concerned, there hasn't been much attention paid to the markings, shape or size of the human ear.

There was a small flutter of interest in the ear in 1974, when a paper in the *New England Journal of Medicine* claimed that people who had a diagonal crease in their earlobe were more likely to suffer heart disease.

So I was extremely delighted to read, in the 1995 Christmas Issue of the *British Medical Journal*, a paper by James Heathcote discussing the relationship between the size of human ears and age. He did the scientific thing and ignored the question of "Why do your ears get bigger?" Instead, he went back to the far more basic question of "Do older people actually have bigger ears?"

First, Heathcote searched the medical literature. To his surprise, he found that nobody had ever tried to answer this question, at least as far as adults were concerned. There was a German study that looked at 1271 children and found that their ears did get bigger as they grew older (which is not a real surprise), but there were no studies on adults.

Wha'd ya call me?

The English Ear Experiment Begins

Shortly after, four English GPs began measuring the length of the left ear of all their patients over 30 years of age. They accepted people of either sex, and of any racial group. Eventually, their study looked at 206 people, aged between 30 and 93. The ear lengths ranged between a delicate 52 mm (2 in.) and a whopping great 84 mm (about 3.3 in.).

In their graph, there were data points all over the place. But there was a general trend that people's ears did get bigger with age — at the rate of 0.22 mm (0.009 in.) per year.

Japanese Scientists Respond

Soon after, some Japanese doctors (Yasuhiro Asai, Manabu Yoshimura, Naoki Nago and Takashi Yamada) looked at 400 patients aged 20 and up. Now, of course, taller people have bigger ears, and the British hadn't taken account of this.

The Japanese then did a bit of clever massaging of the statistics. They found that, even allowing for the fact that taller people have bigger ears, older people still had bigger ears than younger people.

Belgium and Holland Retaliate

By this time, national pride was at stake. Surprisingly, it was two doctors from Belgium and Holland who counterattacked with some even more clever statistics, which they published in the 1996 Christmas issue of *The British Medical Journal*. Jos Verhulst, a researcher from the Louis Bolk Institute in Driebergen, the Netherlands, and Patrick Onghena, Associate Professor in Educational Statistics from the Faculty of Psychology and Educational Sciences at the Katholieke Universiteit in Leuven, Belgium, analysed once again the original data that had started off the whole debate.

They worked out the average ear length at different ages.

The doctors found that, in the sample of 206 English patients, there was a strange seven-year cycle in ear growth. When the people in the study were 35, 42, 49 and 56 years old, their ears grew (on average) at around 1.5 mm (0.06 in.) per year. But in the years in the middle of the seven-year cycle, the ears actually shrank! This seems ridiculous.

Do Ears Grow Bigger?

At this stage, I began to think that they had used very powerful statistical engines on a very small sample size and come up with complete garbage. For example, they didn't mention how many in the sample were women who wore heavy earrings!

But Professor Kay-Tee Khaw had an interesting suggestion. Perhaps, he wrote, *"Big ears predict survival. Men with smaller ears may die selectively at younger ages. Ear size or pattern, or both, may be a marker of some biological process related to health."*

WAR OF JENKINS' EAR

Did you know that an ear triggered nine years of warfare?

The ear of Captain Robert Jenkins was supposedly cut off by Spanish pirates. In 1738, he held up his severed ear before a Committee of the House of Commons, claiming that Spanish Coast Guards had cut it off in April 1731 in the West Indies. After they had boarded his ship and cut off his ear, they pillaged the ship and set it adrift.

English public opinion was already very much against the Spanish. This incident led to the famous "War of Jenkins' Ear" between Spain and Great Britain. This war began in October 1739, but very quickly turned into the War of Austrian Succession, which lasted from 1740 to 1748.

R. M. Hardisty, a professor of haematology in London, came up with an alternative to the "your-ears-grow-longer-as-you-get-older" theory. He wrote in the *British Medical Journal*: "*Have the senior citizens in the sample had big ears all their adult lives, and will the younger members keep their smaller ones? If so, what . . . might have been responsible? I wonder whether there has been a steady decline in the boxing or scrubbing of children's ears, or whether big ears are simply another result of passive smoking.*"

One thing is clear — on average, older people do have longer ears.

The other thing that is certain is that this debate is definitely not over.

I can hardly wait for more Christmas issues of the *British Medical Journal*. But until then, like the Chinese emperors of old, I'll keep my ear to the ground!

EARWAX

Earwax protects your outer ear. It's also called "cerumen". It's very sticky, so that many foreign objects and invaders (such as dust, pollen, bacteria, fungi, etc.) stick to it. Earwax also contains enzymes (lysozymes) that can break down the cell walls of invading bacteria.

The colour and texture of earwax depends on your race. Mongolians have a grey, dry earwax, while most white people have a honey-coloured, wet earwax.

The tendency to have moist earwax is inherited as a dominant trait, while dry earwax follows a recessive inheritance.

REFERENCES

Yasuhiro Asai et al, "Correlation of Ear Length with Age in Japan", *British Medical Journal*, Vol. 313, 2 March 1996, p 582.

R.M. Hardisty, "Lifelong Follow Up Study of Young People is Needed", *British Medical Journal*, Vol. 313, 2 March 1996, p 582.

James A. Heathcote, "Why Do Old Men Have Big Ears?", *British Medical Journal*, 23–30 December 1995, p 1668.

Kay-Tee Khaw et al, "The Chinese Believe That Long Ears Predict Longevity . . . and That Thick Ears Signify Greater Wealth", *British Medical Journal*, Vol. 313, 2 March 1996, p 582.

"Prominent Ears: A Celtic Perspective", *British Medical Journal*, Vol. 313, 2 March 1996, p 1715.

Jos Verhulst and Patrick Onghena, "Circaseptennial Rhythm in Ear Growth", *British Medical Journal*, 21–28 December 1996, pp 1597–1598.

NOSTRADAMUS PREDICTS THE PAST?

If you've ever spent time poring over the astrology columns in the tabloid newspapers, then you would have heard of Nostradamus. He was a medical doctor who lived in France in the 1500s. He wrote 942 four-line verses, which supposedly predict the entire future of the human race from the mid-1500s until the year 3797.

Every time you get some type of natural disaster, such as an earthquake, or even a natural, regular, non-threatening event like an eclipse, people start crawling out from under stones, claiming either that it means the End of the World As We Know It, or else that Nostradamus predicted that this would be the End of the World. Nostradamus supposedly predicted the rise of Napoleon, Hitler and Stalin, the fall of the czars in Russia, all the major wars of history, the assassination of President Kennedy and probably even disco.

Apparently, Nostradamus also predicted that the King of the West would be known as an adulterer. He could have been referring to more than one American president.

The whole Nostradamus Prediction Industry seems overblown.

The note on the table reads (upside down):

A mighty thunder came from
all around with smoke and fire
silver balls turned over man +
woman hearts beating,
from them came lycra,
white slax and overpriced
wine coolers.

Who Was Nostradamus?

Nostradamus, also called Michel de Notredame, was born in 1503 and died in 1566. He studied medicine and began medical practice in 1529 in Agen. Around 1534, Nostradamus married a beautiful and wealthy woman, who bore him two children. He moved to the town of Salon (where he is buried) in 1544.

It was in Salon that he got a lot of credit for his courage and innovation in combating the plague of 1546–1547. His wife and children died of the plague. In 1554, he married a wealthy widow, Anne Ponsart Gemelle, who bore him another six children. He was appointed Physician-in-Ordinary to Charles IX in 1560.

The Predictions

It was immediately after the plague that Nostradamus began making prophecies.

He would "see" the future by "scrying".

Scrying is a method of divination that goes back thousands of years, but you can try it for yourself at home. Place a bowl of water on a brass tripod at nighttime, and then stare into the water until you start getting some decent hallucinations, or visions.

Nostradamus wrote up his visions using a rhyming four-line poem called a "quatrain". Altogether, he published 942 quatrains or predictions. He published them in 1555 in a book called *Centuries*. They were grouped into nine sets (or "Centuries") of 100 quatrains, and one set of 42. It's the seventh set that has only 42 quatrains.

The Background

France, in 1555, was a turbulent and interesting place — perhaps too interesting — with plagues, high death rates, very poor medical care, no antibiotics, no surgical anaesthesia, and lots of nearby unfriendly nations. The Moors had just been pushed back out of Spain about half a century earlier. The Ottoman Empire was invading across from Turkey and was only about 80 km (50 miles) from Vienna.

In 16th-century Europe, the censors could be pretty heavy-handed. There was no such thing as freedom of the press. It was one thing to make some extra money by writing a book of predictions, but you didn't want to get burnt at the stake for witchcraft. So being obscure and hard to understand was an advantage.

Nostradamus' quatrains were already pretty obscure because they were poems, which had to rhyme and follow a beat of 10 syllables for each line. He deliberately did not order them sequentially, but published them in a random order. Finally,

he wrote very cryptically and ambiguously, covering his tracks so that the censors wouldn't come and burn him at the stake.

His Book Hits the Streets

Nostradamus' book sold fairly well and brought in some income, even though the critics panned it. Videl, in 1558, had a very low opinion of this "*lunatic brainless fool*", and Fulke, in 1560, was highly critical of his "*useless astrological predictions*".

But the bad reviews made no difference. Nostradamus has always had his fans over the centuries. They reprinted and re-interpreted his prophecies.

His book was reprinted in 1568 in Lyons, and a few years later in 1589 there was the Paris edition of *Les Prophetis de M. Michel Nostradamus*. And then in 1605, Pierre Duruau published another edition in Troyes.

Nostradamus never wrote any specific predictions; they are deliberately very vague. It was his fans who later re-interpreted his work to give us the specific predictions.

Henry Roberts kept the fans happy with his 1947 book on Nostradamus. Edgar Leone wrote a massive and definitive book about Nostradamus' predictions in 1961, *Nostradamus: Life and Literature*. He based his book on the 1605 Troyes edition. But Leone's book didn't really set the bookshop cash registers humming.

Nostradamus Gets Really Popular

The current wave of interest in Nostradamus came about because of the work of Ericka Cheetham, who died in 1998.

She had studied Ancient French and Ancient Provençal at university. Cheetham had her own personal copy of a 1568 edition of Nostradamus' work and was also able to get access to other early editions of Nostradamus' writings at the British Museum. So, spurred on by a commission from an American publisher to write yet another Nostradamus prediction book, she started work in 1969.

Her first Nostradamus book, the *Prophecies of Nostradamus*, was published in 1971. Her next book was the *Further Prophecies of Nostradamus* in 1985, and she followed up with the *Final Prophecies of Nostradamus* in 1990 (which was basically a revised version of her original 1971 book). People bought them like there was no tomorrow.

Post- or Predictions?

The Nostradamus Industry tells us that many of his predictions have come true over the ages. For example, Nostradamus is supposed to have predicted the American Revolution and Civil War, as well as every other major event in history from the First and Second World Wars, up to air and space travel, and even the landings on the Moon. But these predictions were first interpreted *after* the events had happened.

After all, it is much easier to "predict" the past than the future. The only "sight" that is perfect is "hindsight".

But the Nostradamus fans always support their argument with two very important predictions that were supposedly made *before* the events happened. However, the first one was wrong, and the second one was written up after the event happened.

First Prediction — Henri II

Nostradamus is said to have predicted the death of the French king Henri II in a tournament in 1559. If we go to the first of the 10 Centuries of quatrains, and look at the 35th one, we read the following four lines:

> The young Lion will overcome the old one,
> In martial field by a single duel.
> In a golden cage he shall put out his eye.
> Two wounds from one, and then he shall
>   die a cruel death.

Well, yes, Henri II did die in a tournament, but forget about the accuracy of "*young Lion and old one*" — the contestants were about the same age.

The mention of the "*Lion*"? The lion is a symbol that has been used by various royal families, but unfortunately, neither Henri II nor his opponent used the lion as a symbol.

> In a golden cage he shall put out his eye.
> Two wounds from one,

The helmet that Henri II used was neither gilded nor gold, and there was only one wound — and it did not affect either of his eyes.

But the Nostradamus Industry isn't fussed by minor inconvenient details. It has turned this very inaccurate quatrain into a spot-on, forensically accurate description of the death of Henri II, and Nostradamus has been hailed a successful seer.

Second Prediction — Sixtus V

The next often quoted prediction is that Nostradamus named a quiet young monk as destined to be a future Pope.

The young monk was Felice Paretti. The story goes that Nostradamus met him by accident in the street, saw the future, knelt down and called the young monk "His Holiness". It is true that Felice did eventually become Pope Sixtus, but this holy event happened in 1585, 19 years after Nostradamus died. The first mention we have of this meeting-in-the-street surfaced in 1590 — 24 years after Nostradamus died, and five years after Felice was made Pope.

This story was totally unknown while Nostradamus was still alive. So it seems this prediction was also made *after* the event.

(By the way, in the late 1550s, Paretti held the position of Inquistor-General in Venice. He was actually recalled because he was too enthusiastic in his work, which is a remarkable and horrifying achievement for an Inquistor in the 16th century!)

Nostradamus Predicted Saddam Hussein?

Some people link Saddam Hussein and the 1991 Gulf War to a quatrain where Nostradamus specifically mentions the Antichrist "Mabus". This is in quatrain 2:62, or the 62nd quatrain in the second Century. It reads:

> *Mabus will soon die, and then will come*
> *A horrible slaughter of people and animals*
> *And once vengeance is revealed coming from*
> *   a hundred lands*
> *Thirst and famine when the comet will pass.*

The Nostradamus Industry tells us that when you spell "Mabus" backwards, you get "Saddam". But when I spell it backwards, I get "S-u-b-a-m", which is different from "S-a-d-d-a-m".

Dozens of Nostradamus fans used this quatrain to predict the Gulf War — *after* the Gulf War happened. But not one Nostradamus fan interpreted this quatrain *before* the war happened. And anyway, this quatrain says that Mabus is killed early on in the conflict, but Saddam Hussein is still alive and kicking.

Nostradamus also refers to conflicts happening after a bearded comet appears in the skies. If the comet doesn't have a beard (nowadays called a "tail"), then it's not a comet, and unfortunately, we're always having wars.

Two Quatrains

You can see how hard it is to predict the future if you look at two quatrains that specifically mention Mesopotamia (currently known as Iraq). Quatrain 3:61 (the 61st quatrain in the third set or Century) reads:

> *The great troop and cross-bearing sect*
> *Will arise in Mesopotamia*
> *From a nearby river the light will come*
> *Which such a lore/religion will hold for an*
> *   enemy.*

Quatrain 8:70 reads:

> *He will enter, vile, wicked, infamous*
> *Tyrannising over Mesopotamia,*
> *He makes all (his) friends by the adulterine lady*
> *Horrible land, black of physiognomy.*

So let's see what different Nostradamus interpreters made of these quatrains.

Six Predictions

Back in 1941, Lee McCann interpreted these quatrains to mean some sort of trouble with Arabs would occur, some-

THE GREAT ECLIPSE OF
11 AUGUST 1999

Quatrain 10:72 reads:

The year 1999, seventh month,
From the sky will come a great King of Terror:
To bring back to life the great King of the Mongols,
Before and after Mars to reign by good luck.

Thanks to the current distances of the Earth, Moon and Sun, we can have lunar eclipses and total solar eclipses. If you analyse the motions of these bodies, you will see that there can be up to seven eclipses (of the Sun and the Moon) each year — so eclipses are not very rare. However, the Nostradamus Industry sees every eclipse as a Harbinger of Doom.

This quatrain obviously refers to August 1999 (since August is "close" to the seventh month), when there was a total eclipse of the Sun visible across parts of Europe.

The Nostradamus Industry swung into gear, and predictions poured forth.

One said the "Great King of Terror" was obviously Comet Lee (Comet C/1999 H1), which had a huge asteroid hiding in its tail. This prediction then went on to interpret quatrain 3:34:

When the eclipse of the Sun will then be,
The monster will be seen in full day:
Quite otherwise will one interpret it,
High price unguarded: none will have foreseen it.

Obviously, this particular quatrain predicted that the asteroid would become visible for the first time during the eclipse. It would (according to interpretations of other quatrains) smash into the Mid-Atlantic Ridge, open up vast volcanoes, and raise 1-km (1100-yard) tidal waves on all Atlantic coasts.

And, of course, the Nostradamus Industry said this was all backed up by quatrain 6:6:

There will appear towards the North
Not far from Cancer the bearded star:
Susa, Siena, Boeotia, Eretria,
The great one of Rome will die, the night over.

The other major prediction was related to the Cassini probe, which will arrive at Saturn in 2004. It did a mid-course correction on 11 August (same day as the eclipse). This sent it close to the Earth to do a gravitational slingshot manoeuvre on 18 August. The Nostradamus Industry said that the mid-course correction would go tragically wrong, and that Cassini would not miss the Earth. Instead, it would scatter about 33 kg (72 lb) of plutonium over the entire planet.

Neither of these predictions came true.

And there's more.

French fashion designer Paco Rabanne wrote a bestseller, *1999 — The Fire from the Sky*. He predicted that the Russian space station, Mir, would crash on Paris on 11 August. He said that if his prediction did not come true, he would *"forever be silent and work on nothing but fashion"*.

where around Mesopotamia, which would be all over by August 1987. This interpretation doesn't really tell us who, what, where, when or why.

In 1947, Henry Roberts declared that quatrain 3:61 meant "*a great organisation, with some sort of cross as its banner shall emerge in a land between two rivers. Near one of these rivers, some traitors shall give the enemy assistance*". Quatrain 8:70 meant that "*the country near Babylon will be terrorised by a person of the Negro race*".

This is another prediction that is not big on detail.

In 1975, Ericka Cheetham really blew it when she ignored the fact that Mesopotamia is now called Iraq. She thought that Nostradamus was talking about the area between the Seine and Marne rivers in France, and that quatrain 3:61 was predicting Germany's invasion of France in 1940. Quatrain 8:70 meant almost the same, except she turned Mesopotamia into Avignon in France.

In 1980, Jean Charles de Frontbrune thought that these quatrains referred to conflicts in the future between the West and the Arab nations — again, he gives us no hard facts.

In 1981, Rene Noorbergen interpreted these quatrains to mean a conflict between the West and the Arab nations, but he added steroids and threw in the whole world. He saw this as happening immediately after the Earth had been blasted by a meteor, China had launched both nuclear and germ warfare attacks against the West and had invaded Russia, and after China had teamed up with the Arab nations to invade Europe. Somehow, England was to be totally flooded, so the English would leave to invade France, while the Americans and the Russians would decide they weren't going to sit around doing nothing, and would recapture Europe, making the Chinese surrender.

This time we got a lot of specific detail — but hey — none of it came true.

In 1989, Delores Cannon got involved with the Nostradamus Industry. She had an advantage over most of the other Nostradamus interpreters because she used

THE LANGUAGE OF NOSTRADAMUS

When the Nostradamus predictions are hyped by the media, they usually say that they were written in "*a hybrid of archaic Provençal, with Latin, Greek, Italian and English phrases*".

In fact, he wrote his quatrains almost entirely in French. He threw in a few words of Latin and Old French. He used less than 100 non-French words and no English words.

hypnotic regression, and spoke directly with Nostradamus. Her book was called *Conversations with Nostradamus: His Prophecies Explained*. Even though she had the advantage of being able to talk turkey directly with the Big N, "*albeit through hypnosis from several different subjects*", she completely ignored this quatrain because she probably didn't think that anything important would ever happen in Mesopotamia.

None of these Nostradamus Experts rang the bell on the accuracy meter even once. They're very good at "postdictions", but lousy at predictions.

REFERENCES

Almanac of the Uncanny, Reader's Digest, 1995, pp 106–107, 185–186.

Elizabeth Gleick, "The End Is Nigh — Perhaps", *Time*, 12 July 1999.

Allan Lang, "Are Skeptics Really Impressed by Nostradamus?", *The Skeptic*, Vol. 16, No. 1, Autumn 1996, pp 11–12.

Allan Lang, "Nostradamus and the Middle East Crisis", *The Skeptic*, Vol. 10, No. 5, Spring 1990, pp 8–11.

Allan Lang, "Ericka Cheetham 1939–1998", *The Skeptic*, Vol. 18, No. 3, Spring 1998, p 39.

Tim Larimer and Sachiko Sakamaki, "The Doom Machine", *Time*, 5 July 1999.